Diving into the bitstream

This book weaves together the concepts and conditions of IT to offer a contextualized look at one of the most popular, relevant, and promising industries of today. But what distinguishes this book is its focus on the impact of IT on societies, and the responsibilities of IT's creators and users. The author pulls together important, often complex issues from the relationships among information, information technologies, and societal constructs. With its wide array of topics and easy-to-process language and presentation, this book creates a space for a reader to not only learn, but also to evaluate and question the implications of IT's place in society.

This text covers various issues related to organizational/societal social responsibility and information technologies. Social responsibility encompasses such objectives as promoting the well-being and dignity of individuals, of the diverse communities in which we participate, and of society at large, as well as maintaining a profound respect for the environment and for our position as environmental stewards. Topics include ethical, social and cultural issues, including IT-enabled access for the disabled, intellectual property, internet security and privacy, social networking, asynchronous learning, artificial intelligence, and more.

M. Barry Dumas is a Professor of Computer Information Systems at the City University of New York, Baruch College, Zicklin School of Business, USA.

Diving into the bitstream

Information technology meets society in a digital world

M. Barry Dumas

Routledge
Taylor & Francis Group

NEW YORK AND LONDON

First published 2013
by Routledge
711 Third Avenue, New York, NY 10017

Simultaneously published in the UK
by Routledge
2 Park Square, Milton Park, Abingdon, Oxon OX14 4RN

Routledge is an imprint of the Taylor & Francis Group, an informa business

Library of Congress Cataloging in Publication Data
Dumas, M. Barry.
   Diving into the bitstream: information technology meets society in a
   digital world/M. Barry Dumas.
   p. cm.
   Includes bibliographical references and index.
   1. Information technology—Social aspects.   2. Information society.
   I. Title.
   HM851.D856 2012
   303.48'33—dc23
   2012007625

ISBN: 978-0-415-80713-5 (hbk)
ISBN: 978-0-415-80714-2 (pbk)
ISBN: 978-0-203-15327-7 (ebk)

Typeset in Minion Pro, Optima, Corbel and Syntax
by Florence Production Ltd, Stoodleigh, Devon

SFI label applies to text stock

Printed and bound in the United States of America
by Edwards Brothers Malloy, Inc.

For Laura, Dave, Steve, Liz and Clark. You make the future look bright.

Contents

Acknowledgments

I am grateful to the many people who contributed to making this book a reality. John Szilagyi, Publisher—Business & Management, lit the spark and saw the work through to completion. Manjula Raman, Editorial Assistant, Business, Management & Accounting, worked hard on preparing the manuscript for production.

My special thanks go to Dave Dumas, as always, a great person to bounce ideas off.

Numerous reviewers were involved in critiquing this project at all stages of its development. I thank them for their diligence and appreciate their invaluable suggestions.

About the author

M. Barry Dumas is professor of computer information systems at Baruch College, City University of New York. He has been involved in information systems for over 40 years, in roles ranging from programmer to systems developer, analyst to advisor, practitioner to management, and of course, as teacher. As a consultant he emphasizes responsible practice and consideration of societal impacts as the cornerstone of project development. As a teacher, he encourages independent thinking, active involvement, and accountability as the bases for fulfilling professional and private lives.

Professor Dumas received his PhD in computing and quantitative methods, with a minor in marketing, from the Columbia University Graduate School of Business.

Preface

Nationwide, indeed worldwide, there has been a growing awareness of the importance of access to information. Accordingly, information technology (IT), broadly defined, and its role beyond the internal workings of businesses has leapt into the social consciousness.

As it says in Chapter 1, "While many technologies are narrowly focused, the effects of information technology are pervasive and ubiquitous, especially since the shift from analog to digital. As a consequence of its presence, penetration, rapid growth, degree of sophistication, and even of its absence, no society is left untouched."

Recently, we have witnessed IT's significant role in the Middle East uprisings that have led to the overthrow of long-entrenched dictators. Those events dramatically underscored what has been IT's history as a means of making information available, of controlling the flow of information, and of underlying power shifts, all of which now loom large in public awareness and debate, in college and university coursework, in blogs and social networks, and in other online forums.

What distinguish these trends are the impacts of IT on societies. What distinguishes this book is its weaving together of the concepts and conditions of IT, societal influences, and the responsibilities of IT's creators and users, in the context of our digital world.

Every issue has many aspects, just as every IT has multiple uses and impacts, positive, negative, and degrees in between. In that milieu, our unifying theme is pulling together important, often complex matters in the relationships among information, information technologies, and societal constructs. Our underlying issue is where to draw the line between conflicting objectives, usages, and outcomes, if that can be done at all.

Dealing in detail with everything under this umbrella would require an enormous book, which this is not. Neither is it a cursory survey. Instead, we look at major issues in some depth and bring up others as ancillary concerns to amplify the matters at hand.

Along the way, we raise many questions, but rather than supplying definitive answers, we explore the implications of various positions. The idea is to stimulate further thinking in a quest to encourage development of individual perspectives and judgments.

Every chapter ends with an "And so—" section that brings chapter issues together by examining their interrelationships and consequences.

Chapter synopses

1 What's it all about?   In this chapter we introduce issues and areas that are foundations for further consideration in subsequent chapters and bring up several subordinate ideas as well. The remaining chapters focus on selected themes. Topics to which separate chapters are not devoted are either integrated within various chapters or left in Chapter 1 as food for thought.

The brief encounters of this chapter serve to clarify our interpretation of particular subjects, how seemingly straightforward concepts often have underlying complexities, and how some areas interact positively while others conflict.

After a short introduction, we look at the ideas of communications, information, and information technology in a section called "Foundation ideas, condensed". Then those ideas and a variety of associated topics are explored a bit more in a section called "Now let's look a little deeper".

This section describes how we view the meanings of: information; information technology; new, old, and mass media; societies and cultures; the information age; consumer and provider access; free expression, privacy, property, and ownership; protection and security; the military; e-commerce and e-business; education; artificial intelligence; and green technology and sustainability.

The chapter ends with "And so—IT marches on", and we with it, which delves into the evolving role of IT, the pros and cons of various trends, and some important implications of the digital explosion.

2 Information to suit—as you like it   Information is a rather facile term that's used with abandon, but what is really meant by it? We go beyond the simple data–information–knowledge trilogy to seek a fuller understanding. Considered are information sources and such information characteristics as quality, truthfulness, validity, completeness, and accuracy. We also look at biases, intentional and otherwise, selectivity, understanding, and information overload. Examples show how seemingly straightforward data-based statements can be misleading or simply wrong.

Sections include: "Information—what is it?"; "So we come to knowledge"; "Where does information come from?"; "We know you—data collection"; "US politicos and the media wellspring—information sources too"; "Truth or consequences"; "Fact, fiction, and the in-between—evaluating information"; "We are what we believe—selective narrowing"; "Brain bias—a complication"; and "Understanding numbers—reading between the lines".

"And so—we are what we know" considers what it means to have access to high quality information and what it means not to, whether we are thriving because of the huge amount of information available at our fingertips or drowning in the deluge. We also look at information's evil twin, misinformation, and ponder whether or not information can really set us free.

3 Connections—the Internet, the Web, and the others   The Internet and Web, as incredibly powerful movers and societal influencers, warrant a good appreciation of what they are, how they arose, and where they may take us. In some part, this is a rather technical chapter that provides a historical context for current and perhaps future technological developments. But it also treats some of the implications of the technologies and their impacts.

We discuss: the birth of the Internet and its current structure and status; methods of access and access control; how the Web began and how it is evolving; how information flows over the Internet; the new media/old media link; and shifts in viewpoints of the value and virtue of openness.

Sections include: "In the beginning—a little shiny ball"; "A modest start—the ARPANET"; "Slow progress, then boom—the Internet"; "Finding the trees in the forest—the domain name system"; "Sorry, we're all out of addresses—IP revisited"; "Addresses galore and more—IPv6"; "Consumers and providers—the network neutrality debate"; "On the go—bringing the Internet along"; "The World Wide Web"; and "A Web site for any business—or for you".

"And so—wielding power" considers the implications of the two sides of IT's capabilities: to provide a means for moving information around the world and to prevent information from flowing freely. The questions are: Where does power lie? and Who can take advantage of it?

4 That is to say—free expression and privacy   We want to be secure in our being, free from intrusions surreptitious or overt, confident that our privacy is protected. At the same time, we want to be able to seek information and express our opinions without restrictions, fear of reprisal or retribution.

Embodied in the notions, some would call them rights, of free expression and privacy, are potential conflicts—when privacy concerns limit free expression and when free expression intrudes on privacy. Discussions of such possibilities include consideration of the circumstances under which sensible balances might be achieved. Importantly, can disparities be reconciled?

We explore the meanings of free expression and privacy along with their legal and ethical bases. We consider how privacy has eroded in the digital age and how much information about us is collected without our knowledge. We use examples of actual cases and differences of opinion to illustrate potential conflicts and possible resolutions.

Sections include: "The free expression mandate"; "Limits to protection"; "Censorship —the one-way street"; "Privacy complicates the picture"; "IT affects outcomes"; "The whistleblower—risk and reward"; and "Security, yet another complication".

"And so—where do we stand?" Here we revisit the idea of balancing needs for protecting our free speech rights, our privacy, and our security. We look at some of the changes made in that balance over the last decade or two, the tradeoffs we have accepted, and where we may be headed.

5 What's mine is whose?—intellectual property   Intellectual property, derived from general property rights in concept and materialization, has been the subject of increasingly intense debate impelled by IT developments in the digital age. Beyond discussing what intellectual property is and its various forms, we delve into consti-tutional provisions, legal environments, manifestations of the laws, the bright and dark sides of property rights protections, and how digital developments have vastly altered the intellectual property landscape for better and worse. Examples of lawsuits and other legal tussles, including legislative attempts to secure and alter property rights, illustrate the concepts.

Sections include: "Intellectual property—what and why"; "Copyright—a little history"; "The IT effect"; "Legislators try to cope"; "Unlocking content"; "Patent—a little history"; "The IT effect"; "Taking unfair advantage"; "Trademark—a little history"; and "Trademark dilution".

"And so—innovation stimulator or destroyer?" Intellectual property rights are based on the notion that if we want to encourage creativity and innovation, creators and innovators need to be rewarded for their efforts. Therefore they should have protections against usurpation, against unilateral adoption of their ideas and work. That's what copyright and patent laws are purportedly designed to do. The questions are: Have the laws evolved to such a point where they are actually working against that purpose, where creativity and innovation are now discouraged? Are the trends telling us that it's time to change course?

6 You, me, and everyone else—alone together and vice versa   The transition from analog to digital, from local to global, has brought about remarkable changes in the idea of personal relationships. Before the Internet, keeping in touch with a distant friend or relative meant writing a letter or making a landline telephone call. The former was very slow, the latter very expensive, not to mention that you might not reach the person at all. Sending the same message to a number of people meant repeating the process for each one.

We've moved to a time when keeping in touch is as simple as typing an email, posting a tweet, and sending an update through Facebook or LinkedIn. Whether we're aiming at one person or hundreds, it's all done at the same time with hardly any delay, deliverable globally at very little cost. Moreover, mobile technologies have freed us from having to be at a specific location to send or receive information, for direct contact with someone, and for accessing the Internet to enter websites.

These are wonderful abilities, but how have they changed us? The very concept of friendship has undergone a major transformation. The ability to act irresponsibly and hurtfully has grown immensely. The thought of being out of contact, even for a short time, has become a cause for restlessness.

Information sharing, intentional and otherwise, is now the norm. Privacy has been relegated to a minor role. Social networks are thriving. We live with the impression that we are socially surrounded and involved at the same time as we are physically isolated.

This chapter explores those issues along with particulars of information sharing, cell and smart phones, blogs, video sharing sites, aggregator sites, collaborative computing, social networks, and virtual communities.

Sections include: "Relationships"; "Some definitions"; "A little history"; "Aspects of socializing"; "The socialization debates"; "Impacts rising"; and "Upending the social order—a spark and sometimes a flame".

"And so—the many directions of impact" delves into the influences of this brave new world, our changing expectations, our successes and disappointments, aspects of control, the changing nature of organizations, and the pros and cons of instant reach and always on connectivity.

7 Attacks, bit by bit—from nuisance to cyber warfare It seems inevitable that the dark side of human endeavor is never far away. War of one kind or another has been a characteristic of human interaction since the dawn of history. No day passes without many examples of man's inhumanity to man. It is no surprise, then, that IT is being used to annoy, spy on, attack, disable, disrupt, and otherwise impinge on or threaten humanity's institutions and their constituents.

The history of technological development shows that once we start down a path, it's difficult if not impossible to close it off. Especially with regard to IT, the aphorism *you can't put the genie back in the bottle* holds true. It is unlikely that digital assaults escalating to cyber warfare will be an exception.

We explore the possibilities of using IT for both the offensive and defensive sides of the coin. We consider matters ranging from malware attacks on individuals, specific sites, or installations to hacking aimed at disabling vital infrastructure systems, the latter now possible because of their dependence on IT and other computer-based technologies for their functioning.

What are the ramifications of these growing capabilities? Can defenses keep pace with invasions? What responsibilities do we and our governments have? From the simplest of single machine malware to the drama of cyber warfare, hard choices have to be made in striking a balance between security and the freedoms we've come to enjoy.

Sections include: "A brief history of hacking"; "A broad range of activities"; "Inside out—leaky barriers"; "The privacy sphere deflates"; "A big bag of tricks"; "Fighting back"; and "Cyber terrorism and cyber warfare—a rain of bit bombs to attack people, systems, and infrastructure".

"And so—keeping the bad guys at bay". The huge multiplier effect of the Internet together with the rapidly growing capabilities of IT have given us unimagined power to seek and move information around the world. At the same time, that has opened doors into disruptions on a scale previously found only in the realm of fiction. Protecting us and our organizations has spawned a multibillion dollar industry that together with research institutes and government-sponsored programs try to keep one step ahead of the miscreants of the world.

None of the activities of today's malefactors is more chilling than the eerie parallel to the cold war, a period of nearly half a century that began shortly after World War II. That was a time of very high tension, militarily and politically, with the US and its allies on one side and the USSR and its member countries on the other. The terrifying threat was the possibility of a global nuclear conflagration that would for all practical purposes destroy civilization, if not all life on the planet.

Embryonic cyber warfare is on the radar now. There is no doubt that capabilities for carrying it out will develop rapidly, and so we are entering a period analogous to the old cold war, where the threats of actual cyber warfare will be used in a game of blink just as threats of nuclear war once did. One thing that kept the old cold war cold was mutually assured destruction—the ability of one side to strike back and destroy the other even as it was itself being destroyed. Does mutually assured destruction apply to the new cold war, keeping it cold until maturity and sanity end the standoff? Is IT a difference-maker in that arena too?

8 Mind and machine—artificial intelligence   Computer capabilities continue to grow exponentially while cost and size decrease at a similar rate. The faster smaller cheaper mantra hasn't let up yet. Whether or not there are limits to this progression, the question remains as to the roles that computer-based devices of whatever degree of intelligence will play in our lives. Perhaps the ultimate question is: Will we reach a point where what has so far been uniquely human intelligence will be imbued in or even exceeded by machines?

We explore what intelligence means, what great benefits might accrue from significant advances in artificial intelligence, and where caveats may lie. Machines that can access, manipulate, create, utilize, and disperse information can be viewed as boon or doom. Opinions and sentiments grow more intense, emotional, and polarized as talk expands to the potential of artificial intelligence (AI) to reach or supersede the capabilities of human intelligence.

Research is proceeding apace and some remarkable developments have already been achieved. As AI grows more and more powerful, we once again are confronted with a question of limits—Where do we draw the line between what is permissible and what isn't, or do we?

Sections include: "A little history"; "A different view"; "The imitated mind"; "Progress on many fronts"; and "Becoming ordinary".

"And so—master or slave?" Reasoned logical scenarios are being posed in which, in the not too distant future, artificial intelligence will reach a point where machines can learn from their mistakes and experiences, adapt to changing conditions, be comprehending, sentient, have emotions, and even reproduce themselves—in other words, be human-like or even superhuman-like. That puts us in a whole new ballgame.

The scope of scenarios is incredible, ranging from simple intelligent devices to assist with our everyday activities to computers and humans blending into cyborgs with extra-human capabilities. Science fiction? Perhaps, but over and over, the science fiction of yesterday has become the technology of today.

So, we can envision a world wonderful or nightmarish. Who or what will be in control? Will advanced AI machines serve us or will we serve them? Sensible or not, fears have led to thoughts of limiting AI development, the same sort of fears that have been used as arguments to keep other technological advances in their places or stopped altogether.

9 We are what we do—ethics and responsibilities   As digital IT increasingly becomes a part of the world and normal life, ethics and IT are ever more inextricably linked. While ethical considerations have cropped up in previous chapters, this one treats the subject in greater detail, exploring prior discussions more fully and looking at some new ones.

Some ethical questions are easy to resolve. Other seemingly straightforward ones may not be. We look at philosophical bases for ethical decision making and use examples to see how they may apply to issues in the digital age. Some examples:

- Software for manipulating digital images and videos lets us be remarkably creative. It also lets us obfuscate, mislead, and outright deceive. Are the software producers responsible for the uses of their products? Can they incorporate the means for detecting when an image has been modified? Should they?
- It seems simple enough to say that it is unethical to leak private information, but is it always? The revelations of whistleblowers may arise from malice, but they also may properly expose wrongdoing. Once information is leaked, it's a minor matter to make it globally available regardless of its validity, its purpose, or its consequences. A look at some cases illustrates the issues.
- Various means of tracking our Internet use, Web surfing habits, and physical movements are being more widely employed, ostensibly with benign and beneficial intent. Improved searches, targeted advertising, easier associations, and the like may be ethical, or they may constitute invasions of privacy. A difference-maker is whether or not we have knowledge of the practices and can choose whether or not to participate.
- The search engine game is in full force. On one side are businesses that attempt to gain high placement on results lists by outfoxing ranking algorithms. On the other are search service providers that want to keep algorithms under their control, possibly for our benefit but possibly for their own. Is either side acting ethically?
- Though we have made some strides in dealing with toxic e-waste, a lot is still dumped on people who have little choice in handling it. But those exploited workers are not operating on their own—they are employed by people in their own countries who take advantage of their dire straits and profit from their labors. The ethics and responsibilities of the e-waste challenge redound as much to those employers as to the producers and transporters of the material. And then there are governments on all sides—are they complicit?
- Among its detractors, artificial intelligence is denigrated because they do not believe any machine can have such presumably unique human characteristics as a conscience, a sense of responsibility, an intuitive grasp of right and wrong. Is it possible to build moral machines? If we can reach the point of a machine brain having all the capacity of a human brain, will ethical awareness and a conscience naturally follow? Can such a machine learn moral behavior as humans do?

Sections include: "A little background"; "Some major concepts"; "IT enters the picture"; "Intersections"; "Free expression"; "Privacy"; "Intellectual property"; "Artificial intelligence"; "Professional ethics"; and "The green way".

"And so—reaping what we sow". Who benefits from the digital age, the blossoming of IT, the incredible fingertip-ready access to resources, and who has it failed to reach? Is the common good better served if universal access reigns? Is it reasonable to expect the haves to subsidize the have-nots? Can ethical sensibilities and common sense help steer prudent legislation while at the same time serve as a guide to reason that can fill in the gap between lagging legislative action and rapidly advancing technologies? Pondering these questions brings together the issues discussed in the chapter.

10 The future lies ahead—we're off to see . . .   The thrust of this chapter is to pose interesting and stimulating scenarios as natural developments of some current ideas and practices. Rather than being an attempt at prognostication, we seek instead to encourage thinking about future possibilities and their ramifications, posing some what-if developments and alternate directions that may be taken. We believe that this makes for a more useful concluding chapter than a summary of what's already been discussed. Some possibilities:

- Freedom of the press, a long-cherished foundation of democracy and open societies, requires having a press. It used to be that hardly anyone did. Now with blogs, tweets, social network postings, video aggregator sites, and the like, anyone with an Internet connection effectively has a press and can present ideas unfettered. We can choose to exercise that freedom responsibly, or un-encumbered by commitments to accuracy, reliability, and fair representation. What are the implications?

- Newspapers, and to a lesser extent radio and television news broadcasts, once depended on journalists to go beyond simple reporting of events to investigate circumstances, verify facts, follow-up on developments, and take the appropriate amount of time to do so. Now, the immediacy of Internet postings has changed the meaning and practice of journalism and reportage dramatically, shifting a large part of the responsibility for accuracy, completeness, reliability, and credibility from content producers as trustworthy sources to content consumers as evaluators and detectives. Where are we headed?

- Tracking via cookies, key loggers, GPS-enabled smartphones, and the like have been with us for some time. Tiny intelligent motes called smart dust may be taking us to a new dimension of tracking. Are we prepared for the consequences, worrisome or beneficial?

- We gauge people we talk to by interpreting the feelings they convey, meanings beyond their words. Sentiment analysis aims to computerize that judgment. Another branch seeks to do the same for written documents. Does this signal the dawn of a new openness, more accurate assessment of testimony, or another decline in confidentiality, even a way around barring self-incrimination?

- Mind reading has been legitimized by the machines that can read our thoughts. Presently limited in scope and requiring our cooperation and large, expensive equipment, what if we reach a point where our minds can be read without our knowledge or complicity to gather information we hold to ourselves?

- Indications are that Moore's Law, which has reigned for over 50 years, is reaching the end of the line for silicon-based computing. Does that mean that IT advancement also will end or will other methodologies take over? Either way, what can we expect to follow as a consequence?

- Computer intelligence is increasingly able to take the place of skilled professionals. Are large-scale labor shifts on the horizon, accompanied by a changed labor vs. machine battle? Will artificially intelligent computers design their own improved successors and build them with automated machinery?

Sections include: "Some semi-persistent patterns"; "Platform convergence"; "Media convergence"; "Cultural convergence"; "The cloud"; "Revaluing privacy"; "Network neutrality, up or down?"; "The shifting digital divide"; and "Another look at AI".

"And so—the global imperative, keeping up with the chip". The individual and global impacts of information technologies cannot be overlooked and must not be underestimated. Those of us in free societies would like openness to be the global imperative. We look to ever-capable IT to help it come about, just as closed societies look to it to maintain control.

Legislation aimed at keeping a sensible rein on abuses is slow in coming compared with the speed of technological progress. Moreover, one nation's legislation does not have any automatic standing in other nations, nor may it sway their legislation. Unless the course of history changes radically, global international cooperation will remain a utopian vision. What is more likely is that developing technologies will be better tools for preserving conflicting goals and battling to gain the upper edge.

So as we dive into the bitstream, it is up to us to keep our heads above the flow and swim in the right direction. That will determine whether future IT is a boon or a bane.

1

What's it all about?

If you want to understand life, don't think about vibrant, throbbing gels and oozes, think about information technology.[1]

Every technology has its supporters and detractors, its adopters and its repudiators, its benefits and its detriments. Information technology is no exception and just like other information technologies of the past—the telephone, the telegraph, radio, television—the newest have been accompanied by predictions as extreme as leading to the downfall of civilization and putting us on the road to paradise.

While many technologies are narrowly focused, the effects of information technology are pervasive and ubiquitous, especially since the shift from analog to digital. As a consequence of its presence, penetration, rapid growth, degree of sophistication, and even of its absence, no society is left untouched.

It is not possible to cover every aspect of information technology and society in a single book. To give you a notion of what we will explore, this introductory chapter provides synopses of issues and areas that are foundations for further consideration. Separate chapters are devoted to some of these, others are integrated within various chapters, and still others are left in this chapter as food for thought. These are noted in the discussions that follow.

Foundation ideas, condensed

. . . the number one benefit of information technology is that it empowers people to do what they want to do . . . and so in a sense it is all about potential.[2]

To effectively communicate, we must realize that we are all different in the way we perceive the world and use this understanding as a guide to our communication with others. [3]

Can you imagine any social order existing without communication? Can you see yourself living in total isolation, completely cut off from any kind of contact? Not likely. Communication is a basic human need, essential to our social systems and interactions in personal endeavors, the business world, formal and informal associations, political alliances, societies, governmental institutions, and every other social structure.

A social construct is anything that exists by virtue of social interactions, as opposed to objective reality. For example, such things as nations, presidents, money and language do not exist outside of the context of human social behavior. Nevertheless, they exist as integral parts of our social functioning.

(http://www.control-z.com/czp/pgs/soccon.html)

Accordingly, information—the stuff of communication—in its aspects of capture, collection, storage, retrieval, display, delivery, dispersion, and employment, is at the foundation of civilizations and their social constructs. It's no stretch to say that information carries with it the potential to enlighten humanity and shed light on events around the world. So it is not surprising that what is communicated to whom and how, as well as who has access to what, are critical subjects that have long been marked by investigation and controversy.

Surrounding information in all its dimensions is technology—so it's called *information technology* (IT). Most importantly among its other functions, IT provides the means for information to flow. Without flow, information is locked up, inaccessible, and therefore worthless. Information untouched and untouchable is invisible—it might as well not exist.

IT can be thought of in many ways: its availability, reliability, access, and control; the regulations, legislation, and oversight under which it operates; how well it protects our security and privacy; our confidence in the accuracy and completeness of the information it handles; and the ethics and principles of its intended and actual uses. These dimensions, subconsciously or directly considered or ignored, influence the IT choices we make.

Logically, we attempt to choose the best technologies for our purposes. Is best definable, even in that context? There are "10 best" lists for guidance, but they hardly ever agree because any determination depends on the personal judgment of the list makers and their definition of best. Among the many possibilities are: functional superiority; marketplace dominance; simplest to use; most commonly employed; most profitable; least costly; most trouble free; most efficient; most compatible; most attractive; or, more likely, some combination of them.

There is no universal measure, nor does there need to be. The opinions of experts, experienced users, and people we trust can guide us, but in the end we each need to make our own choices. In the final analysis, that usually comes down to what seems most appealing, for whatever personal reasons.

That helps illuminate why some technologies flourish and others founder—success defined by the marketplace. Somewhat paradoxically, technologies that blossom are not always the ones rated most highly, while some that are considered to be top notch fail to survive. Of course, many excellent technologies have triumphed and many that fail to catch on deserve that fate. Still, success also depends on more than being technologically superior—at least as important is being the products of well-run businesses. That often is enough to explain why some highly touted products have been tremendous flops even as others less lauded have gone on to considerable success. Business acumen and market insight can be the key difference-makers.

Complicating the picture is the fact that IT is dynamic, continually changing, undergoing adaptation, alteration, modification, and occasionally spontaneous mutation. Developments are impelled by such drivers of change as business, governmental, and societal demands, whose pressures determine the directions of trends as well as which technologies thrive and which fall by the wayside. This is seldom an orderly process. Jumps, sudden shifts, and quirkiness are common. What becomes the next big thing is often a surprise, even to its creators.

Moreover, while every technology embodies expected benefits and detriments in its design and intended uses, we regularly are surprised by unintended consequences, positive and negative, minimal and monumental, that result from actual uses. Some of these may be far from original purposes, which may themselves become secondary factors. How we prepare for and deal with unexpected outcomes has considerable influence on their scope and impact.

All things considered, predicting the next roaring success or how long it will last, is chancy. Ironically, choosing wisely does not always result in the wise choice. The saving grace is that nothing is forever and, with IT, timelines can be very short.

Now let's look a little deeper

In subsequent chapters we explore a variety of issues and ideas. Before we get to them, it's important to explain what we mean by the terminology we use.

Information It may be hard to believe, but there is no uniformly agreed upon definition for information. Instead, it turns out to be a rather facile term that's used with abandon. Even the well-known triad, data–information–knowledge, is not as straightforwardly defined as it appears to be.

In Chapter 2, "Information to suit—as you like it", we compare definitions and seek a fuller understanding. We consider information sources, quality, truthfulness, validity, completeness, and accuracy. We look at biases, intentional and otherwise, selectivity, understanding, and information overload. Examples show how seemingly clear-cut data-based statements can be misleading or simply wrong.

Information technology We regard IT as being inextricably bound to humankind's communication history. The particular technologies—be they stone chisels, quill pens, printing presses, telegraph networks, telephone systems, computers, digital communications networks, and so on—are the underpinnings of the information they carry, whose creation, storage, display, and distribution they make possible.

To properly explore today's communications scene as well as what may be tomorrow's, it's important to understand something of the IT underlying them, including a look back so we may look forward. But first, what about Information and Communication Technology (ICT)? A term that has become popular in the last few years, ICT is said to go beyond IT to incorporate a variety of communications technologies—radio, television, voice and video, and so on—along with access to communications devices. IT, then, is reduced to a focus on equipment, principally computer-based equipment.

We take IT to be an inclusive and comprehensive term. IT handles information and its manipulation in all forms and formats, so communications technologies naturally fall under the IT umbrella. Furthermore, access always was an IT function. So when we refer to IT, we do so in all its dimensions; we do not make an IT/ICT distinction.

The Internet and the Web   These days, when most of us think about IT, quite often the first thing that pops into our minds is the Internet, that global carrier of the roaring flood of bits that comprises so much of the information we seek and send. The Internet is the infrastructure—computers, from pocket size to mainframes; cabling and wireless distribution systems; switching and routing mechanisms; protocols; software; connection providers; and access methods—that supports the World Wide Web and all its applications.

We go online to shop, to get news, to blog, for entertainment, for advice, and so on. This kind of IT communication means connecting to sites in the World Wide Web, a process very much simplified by browsers, the visual interfaces between us and the commands that actually are executed to do our bidding.

The World Wide Web, or simply the Web, is a mélange of hypertext and data files[4] that are chock full of the information we seek and send. Hypertext files contain links that facilitate navigation within themselves and among each other, allowing us to jump directly from place to place within a file and from file to file or to data files. Special scripts run via the browsers let us fill out forms, place orders, check accounts, and more.

Web sites, the repositories of these files, have a presence on the Internet and use it to transport information to and from us and other sites. We seek particular sites according to what we are searching for or what we are contributing to and the Internet finds them for us through an addressing system that uniquely identifies each site.

The scope of the Internet, which in essence is an interwoven information transportation structure, continues to grow by leaps and bounds. Not only is activity of Web-dependent systems and programs surging, but so is non-Web activity.

Though a major part of our Internet use is Web related, a significant amount is not. Email, VoIP (Internet voice calls), RSS feeds,[5] podcasts, streaming audio, and ftp file transfers, among other applications, all use the Internet but not the Web. Beyond that are many non-Internet information technologies, prominent among which are cellular phone systems, ATM machines, cable and satellite TV, satellite radio, broadcast radio and TV, and their various infrastructures. All in all, it's hard to imagine a day when we do not make use of at least one IT-based service.

Since IT is a main component of this book, it appears in every chapter as part of various discussions. But because of its importance, we devote Chapter 3, "Connections —the Internet, the Web, and the others", to multiple technologies in the IT umbrella. As major components of IT, we pay particular attention to the development and structures of the Internet and the Web. We look at other communications technologies as well and also consider the two sides of the IT coin—moving information and preventing information from moving. The latter also is a subject in Chapter 4, "That is to say—free expression and privacy".

New media, old media, mass media   As a term, new media is at once informative and ambiguous. In broad strokes, media[6] refer to such communications instruments as newspapers, television, movies, video, voice, and the like. Information is supplied by many devices in many media formats. Currently, adding *new* narrows the focus to those based on digital processing, distinguishing them from old media, which are based on non-digital technology.

So the term new media has come to mean the forms and functions of the vehicles that manage digitized, hence computer-based, information of all sorts—numerical, textual, images, animations, audio, video. New media have replaced some old media, have supplemented others, and stand alongside still others. Regardless, new media certainly have become the more widespread and the more impactful.

Mass media, on the other hand, refers to reach and size rather than type of technology, *mass* indicating a large, physically dispersed audience. The old mass media of radio, TV, movies, and print publications have far less scope than new mass media, both in terms of audience locations and sizes. Further, old media content distribution takes a considerable amount of time and is not available on demand. Digital technology and the Internet changed that. Reach is, at least potentially, global; distribution time is seconds or fractions of seconds; and availability is nonstop.

So far, so good, but we know that what is new today is old tomorrow. When humans first began to paint on cave walls, they were using what was that age's new medium to record and preserve thought. Other technologies followed, from inscribed stone tablets, to mechanized printing, to radio and television broadcasts, to audio and video recording, to movie film—all were new media of their times. Just as what we learned from those media was prelude to today's new media, so what we can learn from today's new media is prelude to the next generation of "newer" media.

Let's take a quick look back to the 1960s, when the Sputnik-initiated fever took hold in earnest, particularly in the US and Western countries.[7] That was the impetus for rapid growth in digital processing that took off in the last decade and a half of the twentieth century and fueled developments in computing power and availability. The digital age had begun and new media were soon to follow.

Today when we speak of new media we think not so much about digital, which is implied, but of such developments as social networking, collaborative enablers, peer-to-peer sharing, browsers, blogs, wikis, and the like. The lesson here is that media continue to be separate from, not synonymous with, the content they carry. This brings us back to new media's underlying supporting technologies—computers, networks, the Internet and its famed rider, the World Wide Web—that is IT.

Media and IT, joined at the hip as they are, are inevitably part of discussions in many subsequent chapters.

Societies and cultures

A society is people together with the social, business, or political constructs they create and within which they operate. Societies may be quite small—family, friends; somewhat larger—focused common interest groups; quite large—encompassing nations themselves and national and international organizations; and everything in between. They may be formal, informal, or ad hoc, stable or transient, focused or general, public or governmental. Everybody, by choice and by circumstance, belongs to more than one society.

Societies need information flow to communicate, to function. That means IT. So all societies encompass users of media and technology, hence anyone affected by media and technology—in other words, everyone.

There have been dramatic changes in the way societies function vis-à-vis the media and technologies they employ, how they affect those media and technologies, and how those media and technologies affect them. Social, political, economic, and business impacts, along with ethical and legal issues all intertwine as parties to those effects.

Governments, for better or worse, are part of societal webs as well. The world has witnessed the power of IT to play a vital support role in successful popular uprisings against harsh repressive regimes, some leading to their overthrow and others pushing power mongers to make substantial accommodations to their citizenry that otherwise would not likely be forthcoming. While revolutions begin and succeed because of the hard work and sacrifice of people, IT greatly enhances their ability to rally supporters, coordinate efforts, and make the world aware of their plights and their progress. The speed with which some uprisings of late have prevailed owes a lot to IT. The ensuing cultural transformations are remarkable.

As ever a two-edged sword, the very same IT that helps empower populaces is used by autocrats and oligarchic governments to preserve their power. With IT's tools they can track the movements and Internet use of actual or presumed dissidents, spread false information to discredit and counter opposition efforts, eavesdrop on transmissions to ferret out threats, and keep the general public in the dark by limiting information flow, blocking access to particular sites and even shutting off access altogether. In some countries, individuals and opposition members have been successful in getting around these information walls; in others, governments have, at least so far, kept the upper hand.

IT is a boon and a bane in other ways too. On the one hand, the outreach and connectivity we now have is enormous. We can create and participate in many societies, from tiny to international. We can quite readily pursue and establish friendships and business contacts, and interact with special interest groups. We can spread our own take and opinions on matters and events from the most trivial to the globally relevant. Most importantly, distance, cost, and time are no longer obstacles.

On the other hand, social isolation is deemed to be on the rise. The quality and even meaning of friendship has been called into question. We spend less in-person time with family and friends. We play online games with minions around the world but remain physically alone. We hear claims that our ability to jump from site to site, activity to activity, and our proclivity to do so, is contributing to a rise in shortened attention spans and disorders. Internet addiction has become yet another damaging consequence.

In the end, it is informative to think of IT as a tool set. How we use that set is more of a function of who we are in the first place than how the tools work. The difference

IT makes is by functioning as a powerful enabler. But it is we who must decide how to use that power and take some responsibility for our choices.

While the impacts of IT appear throughout the book, Chapter 6, "You, me, and everyone else—alone together and vice versa" looks more closely at social consequences.

The information age   History is compartmentalized by naming ages. Chronicling eras can promote understanding, explain the march of progress, illuminate trends. Now we hear that we are in the information age, implying that at this juncture information is paramount. There is no doubt that it certainly is. The sheer amount of information readily available is unprecedented. Yet the importance of information is not unique to the present—it was a key element throughout history, age-labeled or not.

Think of primitive cultures. Information was vital to their survival. Members of hunter/gatherer tribes had to know how to hunt, where to find prey, how to prepare and preserve their bounty. Farmers in agrarian societies needed to understand what to plant, when, where, and how, and harvesting and processing techniques. Successful traders were the ones who knew where the safe, navigable routes and fruitful markets for their merchandise were.

Early civilizations relied on oral traditions to preserve and pass on the lore of the tribe. As crude forms of writing came to be, communications technologies developed gradually. Now we're experiencing a burst of progress. But always, information of one kind or another was crucial to societies and to success.

Reflect on the maxim *knowledge is power*. Information is the basis for knowledge, so how and whether or not power spreads to the general populace depends on how readily available information is. That's why implicit in the maxim is that power accrues only if there is access to information, an understanding of how to use it, and the ability to follow through. Political dimensions are often a considerable part of the picture, especially to keep control of power.

There is nothing new in this. Tribal leaders and a very few trusted associates were the keepers and wielders of knowledge through oral traditions that carried the important information and history of their tribes and gave the leaders their status. Other members were not privy to that knowledge and so had little power. For some time after writing took a tenuous hold, few people could read or write. Those who could, not surprisingly, were the rulers, institutional and religious, and their principal cohorts.

As literacy and access to information increased, the more open societies were characterized by greater power sharing. Their leaders, subjected to countervailing forces from an enlightened population, found it tough going to be dictatorial. Totalitarianism could not hold sway where information flowed freely. That is why rulers of closed societies spent so much energy keeping the populace in the dark, dispensing only the information they wanted known and clamping down on free flow.

That continues today. As never before, the enormous amount of information readily available to vastly more people is the defining characteristic of our current information age and what makes it historically unique. For better or worse, digital IT is the big difference-maker, whether used for free flow of information, to overcome barriers to information access, or to construct and control those barriers.

We leave this brief historical journey as food for thought, though we do come back to it from time to time. As you read the rest of the book, keep in mind how information characteristics and IT have evolved and will continue to do so.

Consumer access   Our ability to locate and retrieve information or to make it available to others is an access issue. One question is, access how? Focusing on digital information, access often implies entry to the Internet and through it to the Web. So first we need to be able to connect to the Internet and then to be able to reach the sites where the information resides or we want to send information.

That leads to the next question, access by whom? We frequently hear that Internet access is global, as demonstrated by our ability to reach sites at far removed locations worldwide. But global access is not the same as universal access.

A surprising number of people have no ready access, being subject to tight controls on what may be reached and what is out of bounds. Quite apart from access control, and considerably more startling, is the huge number of people who have very limited or no access at all even in countries where access is ostensibly open—commonly called the *digital divide.*

A look at a few statistics[8] is revealing. Asia has over 825 million Internet users, the most in the world. It also has a population of over 3.8 billion and so is only sixth of seven world regions[9] in penetration (percentage of the population who are users), with just 21.5 percent. Africa, with only about 111 million users in a population of over a billion, lags far behind, ranking last in penetration at under 11 percent.

Even North America, which has about 266 million users in a population of more than 344 million and ranks first in penetration at almost 77.5 percent, still has a substantial number of residents who are not online. Penetration is growing worldwide, but we can see that we are still far from achieving universal access and usage.

The third question is, access to what? This is not an all or nothing issue. We cannot logically or reasonably expect full openness; that any website, every page, and each information repository is or should be accessible to anyone at any time. There are good reasons for keeping some information off-limits to those who have no business seeing it—particular government and military data; certain personally identifiable information; business strategies; trade secrets; and so on.

But there is a sharp contrast between those countries where access to a broad range of information and to a great many Web sites and data sources is severely curtailed and those where access can be considered factually and functionally open. If we overlay constricted access to the penetration statistics of particular countries and regions, actual access falls dramatically.

The conclusion is that what we glibly call global communications and universal access is far from either. This issue receives a closer look in Chapters 3 and 4.

Provider access   Content and application providers cover a broad spectrum— individuals commenting on queries or issues, business startups, small firms, large firms, global firms, news organizations, member sites, bloggers, political entities, governments, NGOs,[10] religious organizations, social groups, solo and company-employed programmers and developers—the list goes on. Surrounding all of them is a fierce debate as to quality of access. That is, should every providing person, group, and business—every content and application provider—be assured of equal treatment in the networks that comprise the Internet?

The general term for equal treatment is *network neutrality* (short form *net neutrality*). In a nutshell, those in favor claim that without it some content providers, particularly those with the wherewithal, could make deals to secure higher speeds and preferential treatment to the detriment of others. Further, Internet service providers that also supply content could arbitrarily favor their own, an unfair advantage over those that must contract with service providers to distribute content. Without a level playing field, they conclude, innovation will be stifled.

Those in opposition claim that differentiation is the natural order of things. With strict neutrality, entities that make significantly greater network demands would be doing so at the expense of those that do not. If service providers can't manage network services according to usage, then those that need little capacity would pay for more than they are receiving while heavy users would pay for less—in essence, light users subsidizing heavy users, possibly even their own competitors. Furthermore, system owners would not be able to generate the revenues they need to maintain and expand their networks. Neutrality, they conclude, will jeopardize continued growth and enhancement of network services to the detriment of all.

As yet, there are no universal regulations or even local agreements as to whether or not net neutrality should be mandated by law. But the debate is heating up again, with some very big players maneuvering more openly than before.

Net neutrality receives attention in Chapters 3 and 10.

Free expression, privacy, property, ownership   Free expression, the ability to articulate what we will when and where we will, is not without limits. Likewise, privacy, the right to withhold information we wish to keep confidential, is not absolute. Furthermore, there are times when the two conflict, when we have to disclose what we otherwise would choose not to or refrain from publicizing information we would like to reveal.

There are, of course, criminal means of invading privacy. Conversely, legal activity does not necessarily mean appropriate activity. There are questionable practices and extralegal activities that, at least for the time being, manage to skirt the edges of legality: "news" sites that actually are disguised sales pitches; "free" offers that tie unsuspecting subscribers to expensive contracts; contests and giveaways whose purpose is to create contact lists that are sold for considerable sums unbeknownst to the victims; breaking news stories rife with inaccuracies; inflammatory or derogatory claims and accusations made up of whole cloth.

There is nothing new about these practices, but the Internet and cellular technology have made carrying them out much cheaper, simpler, and harder to stop or overturn. Moreover, the same IT that makes them possible also makes it more difficult to prosecute the perpetrators if they do stray into the illegal, in part because they can operate anonymously and by subterfuge, and be anywhere in the world.

Many companies and sites leave us unaware that information about us is automatically collected without notifying us, requesting or receiving our permission, or informing us about what is done with that information. Amazingly, we may not have access to the very information about us that these entities collect. To whom does that information belong? Is our privacy being invaded? Where does responsibility lie?

Many sites routinely enroll us in particular services if we don't specifically opt out of them. Their default privacy settings give them broad permission to collect and use our personal information as they see fit. Opt in, by which we choose the services we want and the information that can be collected and exclude the rest, gives us more direct control. It's not a stretch to believe that businesses and our desires are often polar opposites in the opt-out versus opt-in question. It is interesting to note that in most European countries opt in is typically the default, while in the US, it's opt out.

We may grant sites permission to collect and use information by assenting to a terms of service (TOS) agreement, an end-user license agreement (EULA), and similar documents. These detail the rights and usages of the provider that we consent to simply by clicking an "accept" check box. We agree to the terms to be able to log onto a particular site, make purchases online, download files, or install software, but do we know what we are agreeing to? Is it unreasonable to expect that we should be well informed before we assent? That the typical agreement is multipage, densely written, and confusing, quite commonly means that few read the magnum opus before accepting the terms.

This brings disclosure into the limelight. Rarely do we fully know what we are agreeing to. Instead, we assume (or hope) we haven't assented to anything inimical. Is it the site's responsibility to make these documents shorter and more comprehensible, or ours to read them no matter how unintelligible they are to most of us? Is it our responsibility to become informed even if it means slogging through the legalese swamp, or is it the responsibility of the sites to at least provide clear, simply worded explanations of what we're agreeing to, what privacy and control we're giving up, what choices we have and how to easily make them?

Ready access combined with the ease of copying and distributing digital material raises another issue—who owns our information and our intellectual property? The adage *information wants to be free* is a populist sentiment not necessarily in concert with many information creators and providers, the sources of intellectual property.

Existing protections, mainly from copyright and patent law, do not address ownership issues in ways that are fully consonant with digital capabilities. Legislation advances much more slowly than technology. As a result, laws dealing with copyrights and patents, for example, are out of sync with IT capabilities. By the time the laws have caught up with today's technologies, we will be using tomorrow's.

There's also an enforcement lag. A newly minted law is not instantly implemented, nor are the authorities immediately up to speed on the law's requirements and how to police them. As a result, enforcers and prosecutors often must rely on legislation that is inconsistent with, or at best marginally relevant to the issues at hand.

These concerns are central to many of the issues involving IT and society and so are treated directly in Chapter 4. Other aspects appear in Chapter 5, "What's mine is whose?—intellectual property", and in Chapter 9, "We are what we do—ethics and responsibilities".

Protection and security   Security has to do with protection from intrusion, misappropriation, and theft. Criminal activity—illegal downloading and distribution of copyrighted material, loosing malware, breaking into information repositories—is a burgeoning problem, one that consumes considerable resources to counter.

The Internet has greatly facilitated an upsurge in the difficulties of protecting systems and sites from intrusions that range from annoying to destructive. As individuals, we are bombarded with malware of all sorts, putting our surfing habits, our account numbers, our passwords, our contact lists, our expressed thoughts, and even our private posts and messages at risk of exposure, capture, and alteration. The same intrusions plague businesses as well. Add to the list hacking into corporate servers to retrieve or destroy private and corporate information, and bringing down sites by denial of service and other disruptive attacks.

An entire multibillion dollar industry has sprung up in response, producing programs and devices to catch attacks before they take effect and to remove malware once it is imbedded in our machines and networks. There is no end to this battle, with the bad guys constantly looking to stay one step ahead of the good guys, who most often must proceed reactively.

Over the years, intrusions have expanded in scope from individual machines to corporate systems, from wired telephones to wireless communications, from desktop hard disks to enormous data centers. These incursions travel over the same networks that carry legitimate information and give us access to all sorts of sites and repositories. The Internet itself, after all, is an inherently unprotected bit transportation system, a free-for-all where anyone with proper knowledge and self-created or easily downloaded software can wreak havoc at all levels. If there is a connection, it can be exploited.

This escalation has reached into the local and national infrastructure systems on which we all depend: electrical distribution, water supply, transportation, government operations, military systems—you name it. All depend on IT to function and so are, to varying degrees, vulnerable and enticing objects of attacks—yet another unintended consequence of the digital age. This has spawned the term *cyber warfare*, the use of the Internet as a vehicle to attack a country's communications networks and infrastructures.

Cyber warfare is so different from conventional warfare, where opposing sides confront each other physically and directly, that the means of attack are outpacing the means of defense. Cyber attacks can be launched from anywhere in the world where there is an Internet connection, whether by governments, by cabals, or by individuals.

Antagonists can hide behind false trails, can easily move their operations to different locations, and can use different paths for their assaults. They may be carrying out attacks on other countries or factions within or outside their own countries. They may have the support of their countries, be moving to bring down their countries' leaders, be acting as rogues for their own nefarious purposes or just to see if they can do it.

To add to the picture, compared to conventional military action, the cost of mounting a cyber attack is miniscule, but since it can come from anywhere at any time and by a great variety of means, the costs of defending against it are huge.

Chapter 7, "Attacks, bit by bit—from nuisance to cyber warfare is devoted to these topics".

The military   Now as ever before, information is a dominating factor on the battlefield as well as being vital to homeland security. But now as never before, IT has transformed the way militaries and security agencies operate.

Military uses of IT range from acquiring real-time battlefield information to running the business-like operations of procurement, inventory management, and logistics. Military IT enables such diverse functions as remote operations of drones, troop movement monitoring, enemy surveillance, skirmish response coordination, missile guidance, pilot training, materiel distribution, strategic planning, and war games.

Beyond being a very heavy user of IT, the military is also a powerful supporter of IT development via project funding. While some of those projects are singularly specific to military and security needs, many of them find substantial civilian use.

It was the US Department of Defense that created DARPA, the Defense Advanced Research Projects Agency, in 1958, specifically for funding research. In what could easily be regarded as its most significant project and spectacular societal contribution, DARPA's first major project funded research and development of the ARPANET, the precursor of today's Internet. The military has continued to support IT development ever since.

Because of its size, urgent needs, and vast budgets, the military has always played a large role in IT development. We highlight some of those impacts in several chapters.

The e-biz buzz—e-commerce and e-business   There's a difference between e-business and e-commerce. E-commerce is the electronic exchange of items of value—goods, services, money—for other items of value; that is, traditional commerce operationalized electronically. E-commerce is one of the fastest growing sectors of the economy. Even the most traditional businesses have added e-commerce to their operations.

E-business is everything needed to carry out e-commerce—the Web site and support for the entire e-commerce operation. That includes the back office functions of transaction processing, payment, order tracking and delivery, site management, communications connections, the people the business needs to engage in e-commerce, and technical support for all those processes.

Recently, the two e-terms have become confused with each other, with references to e-commerce when e-business is meant and vice versa. This is an example of how meaningful distinctions become muddied as misused terminology is picked up and continually repeated. Now the two terms are commonly treated as synonyms.

In any event, we can put them on the same page by noting that since e-commerce requires electronic communications, IT is at the heart of e-business and therefore is an enabler of e-commerce.

E-commerce has had a dramatic impact on the way business is conducted. It began as an alternative to mail and phone orders from solicitations and catalogs. Since then, e-commerce has expanded to embrace every form of commerce from interpersonal barter and exchange to online complements of physical enterprises, banking, stock trading, investing, and strictly online businesses with no physical (brick and mortar) presence at all.

E-commerce has grown to incorporate exchanges between individuals and the operations of governmental institutions, different from what we usually think of as business. As a result, we now have e-commerce operating in every category of electronic exchange: B2B, B2C, C2B, C2C, G2C, C2G, G2B, B2G, G2G.[11] In short, there are e-commerce analogs to practically every type of traditional commerce as well as those that effectively exist only in the "e" world.

Maintaining a competitive edge, or at least competitive par, pushes traditional enterprises to add e-commerce to their other operations. For them, or for completely online companies, doing business electronically means 24/7 availability from anywhere there is an Internet connection. Even if a Web site does not have full functional capabilities, good design creates cost-effective marketing. New solely online enterprises face lower startup and operating costs than do traditional businesses.

As you would suspect, it's not all roses—there are disadvantages too. Web site development and maintenance is critical, an ongoing expense unique to online businesses. An unattractive site will not create a buzz; a poorly designed, difficult to use site will not build a customer base; a site with dated information will quickly lose its following. Online operation also requires real-time data management, security, and appropriate technical support.

An important consideration is controlling site attack vulnerability. Once online, an avenue for hacking exists. That may mean stolen files, damaged files, or site takedown. If the Web site is not operational, not only is business lost but customers may go to other sites and possibly not return.

Comparatively speaking, traditional commerce is relatively well regulated. E-commerce is a harder nut to crack. As is true of privacy, ownership, and security issues, there are situations for which laws and regulations lag behind need. This again is a consequence of the speed with which technology in general and IT in particular changes and advances, in contrast to the much slower pace of legislative action and enforcement.

We leave e-commerce/e-business as food for thought, except for site attacks, which are part of Chapter 7, and search engine optimization, which is discussed in Chapter 9.

Education

At every level, pre-K through university, IT is a challenge that education has had to answer. Since IT development doesn't stop, neither does the issue. It's not just teaching about IT, the very nature of education is changing as well. At all levels, classes are becoming more interactive, with increasing emphasis on participatory hands-on and collaborative learning, and less on teacher–lecture.

In the higher education sphere, students expect greater access to their professors, not having to wait for a class session or office hours to seek help or discuss ideas. Students doing research projects rely much more on Web sources than on print. While evaluating information sources was always appropriate, ready access to the volume of information on the Internet, much of questionable or unknown provenance, means that more emphasis needs to be placed on understanding how to judge validity, reliability, and accuracy of sources, along with what proper use and plagiarism mean.

There is movement to shift away from increasingly expensive printed textbooks to less expensive and easily customizable digital versions, which also benefit publishers by not having to deal with distribution, under or over production, returns, and the used book market. The flip side is what that means for bookstores, increasingly imperiled entities on the way to becoming an endangered species. And what does it mean to have a "used" digital book—can you sell it, trade it, give it away? For the most part, the answer is no, at least for now.

More and more, students feel entitled to choose not only what they learn but where and when, pressuring institutions to provide more online and fewer in-person courses. For the university, this means developing ways to deliver more effective off-campus anytime education—neo distance learning—for what is now called nontraditional students, the very ones who may epitomize the traditional students of the not too distant future.

Could this trend provide important relief from pressure to build more and more physical classroom space and campus housing for growing student populations? Will it result in a major shift in what education means? Will online courses be as effective as face-to-face student–teacher communication? Will the loss of in-person interaction with classmates and professors change the notion of education from producing erudite well-rounded, culturally immersed men and women to one of content learning in isolation? Will the "college experience," heavily involved with personal interaction, go the way of the dinosaur? It remains to be seen—another intriguing topic with many imbedded issues that go beyond the scope of this book. We leave it as food for thought.

Artificial intelligence

There is no universally accepted definition of intelligence, though there is general agreement that higher order intelligence has three key components: *understanding*—being aware of, interpreting, and making sense of what's going on; *reasoning*—the ability to distinguish and select among interpretations or choices; and *learning*—being able to integrate new information, improve understanding, and take reasoned appropriate action as called for.

This focuses on what is generally considered to be the primacy of humans. Of course, animals can understand their environments, learn to adapt to them, and make choices among actions, but not to the level that characterizes humans.

What about machines? If machines can be made to behave as humans would, we can say that they exhibit intelligence. But since machines are not living organisms, that behavior is called artificial intelligence (AI).[12] Perhaps nowadays we'd call it virtual intelligence instead.

Intelligence is not an all or nothing proposition. There is a wide range of intelligence types and levels among humans, and animals and machines as well. Today's computer-based machines already exhibit an array of intelligences, though they have yet to reach the point where machine behavior is indistinguishable from human behavior. Some say they never will, citing an old saw: *true AI is only twenty years away—the problem is, it's always twenty years away.*

Whether or not that dictum ultimately proves to be true, there is no doubt that increasingly intelligent machines are being developed, a fact that excites many people, frightens others, and moves some to states of denial. Joining debates about limits to machine and human intelligences are controversies revolving around ethical and moral considerations, which take their place alongside the practical functional issues of AI.

Chapter 8 Mind and machine—artificial intelligence explores these issues in some detail. Ethical issues of AI are revisited in Chapter 9.

The green way

Taking a sharp turn away from its formerly popular connotation of envy, green now denotes living and acting in ways that are sustainable and kind to the environment. All of us, as individuals and as members of societies and social constructs, bear some responsibility for our impact on the planet, as do businesses and governments of all stripes. We meet this subject at the intersection of green and IT.

Green IT and its companion green computing focus on life cycle—creation, use, disposal. The last, the so-called e-waste issue, is a mounting problem too often "solved" on the backs of the most vulnerable people living in repressive or extremely economically divided countries. Facing a constant and often losing struggle to earn enough to support a family or even themselves, they perform exceptionally hazardous work for menial wages processing toxic e-waste material without appropriate safeguards, because they have no alternatives.

One way to ameliorate that issue is exemplified by the growing impetus to go beyond the life cycle to what we may call the life *circle*, emphasizing reuse and renewal, minimizing the need for disposal. So we go from creation to use to recycle/renew/reuse and back to creation, with very little after-use disposal leaving the circle.

That requires changing the way equipment is designed and produced in the first place. Emphasis must shift to realistic upgrade paths that keep equipment up to date without requiring wholesale replacement and to designing equipment with more benign component materials that make recycling easier and ultimate disposal less problematic.

To succeed, businesses must consider the bottom line. But green systems do not have to mean putting economic viability at risk. Greening via methods that reduce energy consumption in the production and operation of technology, that ease recycling by minimizing the use of toxic materials, and that increase flexibility so as to delay obsolescence for a business's own equipment, can contribute significantly to viability and profitability.

Consider data centers, major users of electrical power for running vast arrays of data storage and distribution devices and for keeping the facilities cool so that equipment doesn't overheat and fail. Green methods can go a long way to reducing operating costs.

Computers and peripherals contain a great many toxic elements. That makes them harder to recycle and hazardous to the recyclers, leading to non-green practices like dumping and shipping to countries that exploit recycling workers. Designing equipment to use fewer or no toxic components can save money and lives, as well as being kinder to the environment.

It's common practice to leave computers and peripherals turned on whether or not they are being used. This produces significantly greater demand for power generation, with its inherent environmental damage. It also increases business energy costs and decreases time to equipment breakdown, which means increased repair and replacement costs. Simply shutting down equipment not in use, or building-in automatic powering down, can ameliorate these outcomes.

This, of course, isn't the whole story, but it illustrates that a little creative thinking and follow through can go a long way to IT greening. It also reminds us that it is not just manufacturers that are responsible. We also must do our part, by purchasing greener IT, by using it wisely, and ultimately disposing of it responsibly.

We come back to IT greening in Chapter 9.

And so—IT marches on, and we with it

The more elaborate our means of communication, the less we communicate.[13]

Every generation sees technological and societal change from two viewpoints—new possibilities and new perils. Today we hear about the isolating effects of computers, the fallacy of friendships in the absence of personal contact, the diminution of real communication, addiction to online gaming, and other dire consequences. That's the substance of the quote, written over two centuries ago in response to the new IT of the time. Fearful reaction is nothing new.

At the same time, we hear about the wonders of growing collaborative opportunities, efficiently maintaining a broad network of acquaintances, more easily finding support, effortlessly keeping in touch, quickly locating information, readily voicing opinions, and so on. Optimism also is not new. Despite fears, real or imagined, the scale has moved steadily to the upbeat side, which often is the case.

As we become accustomed to new technologies we tend to shift from the negative to a positive view. At some point we don't think about the technologies, we just use them. Of course, that doesn't mean the downsides disappear. Rather, over time, as we get over the excitement of new possibilities and become more realistic about the negatives, we are more likely to view adoption as generally beneficial. Continuing on that arc, we find that the technology moves to the background, used but unnoticed.

Electric utilities[14] function as generators, transmitters, and distributors of electricity. Water utilities manage water collection, storage, processing, and distribution. Gas utilities operate gas collection, storage, and distribution facilities. To perform their work, each of these utilities builds and operates large, complex infrastructures, omnipresent yet invisible.

In many ways, IT has become that sort of operation. IT comprises the components of large, complex infrastructures to collect, store, manage, and distribute information. That sounds a lot like a utility and indeed it is becoming increasingly common to think of IT in that way. Just as the essence of electrical utilities is to provide a flow of electricity, of water utilities to provide a flow of water, and of gas utilities to provide a flow of gas, so the essence of IT is to provide for a flow of information.

We are not particularly mindful of electricity, gas, and water utilities when everything is working as it should. In the same way we don't notice the IT behind the flow of information when it is working as it should. But despite their utility statuses, none of these flows is universally reliable and consistently available. There is no country in the world, even the most developed and technologically advanced, for which IT access and availability are complete, and that's apart from the quality of access and availability where they do exist.

Nevertheless, the implications of the digital explosion—with information of every type either being converted from analog to digital or being created in bits in the first place—are colossal. So think about this:

- Individuals now have the ability to contribute and widely distribute information of all sorts.
- There is sufficient memory capacity right now to store in digital form every piece of information that was ever produced in the entire world, and that can be kept up ad infinitum.
- We can perfectly reproduce any digital file as many times as we want, each copy being indistinguishable from the original.
- We can blend the real and the fictional so well that the difference between them can't be seen in the result.
- The capability exists to digitally alter any collection of bits to change its form or message and yet have it look like it was that way in the first place.
- Carloads of information can be transmitted at astonishingly low cost to practically any location at real-time speeds.
- We can write programs that sift through mountains of data to uncover patterns and meanings that otherwise would remain buried and even to create patterns that are not actually there.

And the pace at which developments appear and their capabilities grow is increasing.

That very pace brings with it a downside. Technology in general and IT in particular is neither everlasting nor guaranteed to be backward compatible. We may have wonderful and valuable information stored on our 8½", 5¼", and 3½" floppy disks, but we no longer have the drives to read them. We may have burned our most treasured images, music, videos, and documents onto CDs and DVDs, but as the representations of those bits slowly and inevitably deteriorate, we lose fragments of our treasures—the death of a thousand cuts. And, of course, it's undeniable that CDs and DVDs themselves will become obsolete, as will all the IT familiar to us now.

Today we have more power and ability in the digital realm than was the stuff of science fiction just a short time ago. This can be fantastic or scary, or is that fantastic and scary? No IT is problem free, so digital nirvana will always be beyond reach. Perhaps most widely impactful, our magnificent networks are just as available to miscreants and malefactors as they are to the ethical, moral, and law-abiding of us.

Less dramatic but no less an outgrowth of the digital explosion is loss of focus. Many believe we can use our IT to multitask—to simultaneously do several things at once. Research has shown, however, that for tasks that use the same parts of our brains we actually are forced to quickly flip back and forth among our "simultaneous" tasks—sequential processing.

This requisite time-slicing of concentration means we can't pay proper attention to everything at once. When we focus on one, the others get short shrift, and even the object of focus gets less attention than it otherwise might. That's why we may bump into a lamppost as we walk down the street while texting. More seriously, that's why we may wander across roadway lanes or crash into something when we engage in a cell phone while driving, or check our email, or read or send a text message. Technologies that were considered to be conveniences, time savers, productivity enhancers, have unintentionally become just the reverse when used unwisely.

Is it out of control, doom and gloom, or the dawning of a spectacular leap ahead? A quick look at the past is revealing. It was back on June 14, 1989, when then US President Ronald Reagan gave a speech at the Guildhall in London in which he said:

> Technology will make it increasingly difficult for the state to control the information its people receive. . . . The Goliath of totalitarianism will be brought down by the David of the microchip.

The certainty, simplicity, and inexorableness of that statement, made nearly a quarter century ago, seemed to overstate the case at the time. Still and all, though its prophecy has yet to be completely fulfilled, recent developments in the Middle East have shown that popular will can indeed rattle overlords and demagogues, and make totalitarian governments vulnerable to overthrow.

The massive public demonstrations that brought this about were underpinned by information technologies non-existent at the time Reagan spoke. While there is little doubt that recent successful uprisings against oppressive regimes are owing to and were accomplished by people, there also is little doubt that the IT they used to organize and coordinate their protests and to publicize their dilemmas, activities, and government responses was crucial to their successes.

In sum

Societally and individually, technologies have effects small and large—on our sense of self, everyday living, the characteristics and qualities of our communities, the distribution of power in our societies, the ability of repressive governments to exert control and of restive populations to overcome their oppression. New technologies may have effects in similar or different ways than their precursors, may overtake old technologies, displace them, encompass them, or work side by side with them.

Change is a two-way street. The ways we communicate change our societies and our societies change the ways we communicate. So it follows that how we employ and deploy IT changes our societies and our societies change how we employ and deploy IT.

As individuals and as participants in our social institutions, we and our leaders bear varying degrees of responsibility over each of the foregoing issues. Together we are responsible to and for the social constructs in which we are formally or informally involved and therefore for the impact, influence, and directions of IT.

But information technologies advance faster than the time we have to figure out the best course to take with them or how to deal with their consequences. Nevertheless, there is good reason to be hopeful. The better we understand how information can be used to affect us and societies positively and negatively, the better we can use IT wisely and have a powerful means for serving our more noble instincts.

So the struggle to keep up with and use technology prudently, to create meaningful rules and procedures, to define and follow proper use, to bar abuse, to protect against exploitation and subjugation, to preserve individual rights, and to take responsibility for our actions, goes on. No one knows for certain where the journey will take us, but we are all going there no matter what.

Notes

1   Richard Dawkins (1941–), Emeritus Fellow of New College, University of Oxford, and evolutionary biologist, as quoted by James Gleick in "What Defines a Meme?" *Smithsonian Magazine*, May 2011.

2   Excerpt from remarks by Steve Ballmer, CEO, Microsoft Corp. at the Alamo Area Community Information System (AACIS) grant announcement, San Antonio, Texas, February 17, 2005. See: http://www.microsoft.com/presspass/exec/steve/2005/02–17aacis.mspx for the full announcement.

3   Anthony (Tony) Robbins (1960–), author and self-help coach, in: http://thinkexist.com/quotation/to_effectively_communicate-we_must_realize_that/222507.html

4   In Web parlance, files also are called pages.

5   RSS (Really Simple Syndication) is a group of formats used to easily publish information over the Internet. Also called Web feeds, RSS documents contain both text and information about authorship.

6   Grammatically speaking, media is the plural of medium. It is becoming common to use media as a singular or collective noun.

7   The first Sputnik was launched in 1957 and was an eye-opener. But it took until the early 1960s for the scramble for technological hegemony to reach a fever pitch. More of this is in Chapter 3.

8   Source: http://www.internetworldstats.com/stats.htm, 2010 data.

9   In alphabetical order: Africa, Asia, Europe, Latin America/Caribbean, Middle East, North America, Oceania/Australia.

10 NGOs (non-governmental organizations), generally focus on particular local interests. Though legitimized by government licensing or registration, they operate within a country independent of government control, even if they receive major government funding. Paradoxically, in some autocratic regimes what are called NGOs actually are government-controlled organizations.

11 In this mélange of acronyms, B stands for business, C for consumer, G for government, and 2 for to.

12 The consensus is that the term *artificial intelligence* was coined by computer and cognitive scientist John McCarthy in his proposal for the 1956 Dartmouth Summer Research Conference. He defined it as "the science and engineering of making intelligent machines."

13 Joseph Priestley (1733–1804).

14 A utility provides a public service.

2

Information to suit—as you like it

What lies at the heart of every living thing is not a fire, not warm breath, not a 'spark of life.' It is information, words, instructions.[1]

Information is a source of learning. But unless it is organized, processed, and available to the right people in a format for decision making, it is a burden, not a benefit.[2]

Information—what is it?

Information discussions often begin with definitions. A hierarchy is common, with data at the bottom, information one level up, and knowledge on top. The short-form distinction between the first two is that information is processed data, or from the opposite perspective, that data are raw facts from which information can be developed. That concise definition can be useful, but it doesn't always coincide with practice.

Since the topic has been addressed for a long time, let's see how others treat the terms. Some combine several aspects of data and information into one and others transpose traditional descriptions. Here are a few definitions of data.

> Data: . . . factual information: information, often in the form of facts or figures obtained from experiments or surveys, used as a basis for making calculations or drawing conclusions.[3]

This definition treats data and factual information as synonyms. That works, but doesn't go far enough. The word data does not imply validity. We know that there are good data and bad data, real data and imagined data, so-called facts that are fallacious—still data all.

On the other hand, suppose we go along with the factual qualifier. Then the question becomes, what is a fact? On the face of it, fact invokes the idea of truth— something is a fact if it is genuine, bona fide, if it has been proven or at least we expect it can be, or if it is objectively descriptive. All those words can get us into trouble; more about that later.

> In computing, data is information that has been translated into a form that is more convenient to move or process.[4]

This reverses the "information is processed data" definition. Here, information comes first and becomes data only after it is processed. Bringing computing into play implies that information is what computers take in and manipulate while data is what they spit out.

> Data: "distinct pieces of information, usually formatted in a special way."[5]

So it's the form, not content or validity, that counts, that turns information into data. This also reverses the traditional definition of data.

> Data: "information collected for use."[6]

Another reversed definition of data that makes information the basic unit—it becomes data when it is gathered for some purpose.

And here are some definitions of information.

> Information: "definite knowledge acquired or supplied about something or somebody" and "gathered facts: the collected facts and data about a specific subject."[7]

The first part blends knowledge and information. Does that help? Not much, since we're left with the idea that knowledge and information are one and the same. We prefer to put knowledge on a higher plane than data/information/facts, as we'll see below.

The second part intertwines data and facts, which only need to be collected to become information.

> Information is stimuli that has (*sic*) meaning in some context for its receiver. [8]

This is closer to the common definition of information being processed data and also notes that context counts. Here, however, the term data is narrowed to the special case of stimuli.

> Information: "facts about a situation, person, event, etc."[9]

Information equates to facts, just as we have seen that data does in one of the definitions. Truth (as in facts are things that are true) also is implied.

> Information: "the communication or reception of knowledge or intelligence." and "knowledge obtained from investigation, study, or instruction."[10]

Another inclusion of knowledge, with the added bonus of intelligence, but it reduces information to a transmission role, the sender or receiver of knowledge.

All told, we see that it takes some semantic maneuvering to distinguish between data and information. The problem is, one person's information is another person's data. That is, each of these definitions depends on viewpoint. In essence, there is no substantive differentiation between data and information.

Except in specific contexts, the distinctions are not helpful. Therefore, we use information as an inclusive term, incorporating the ideas of data, facts, suppositions, evaluations, and even opinions, unless there is a compelling reason to do otherwise. This avoids problems of circular definitions and finely sliced but not particularly useful characterizations. It also means that we don't have to talk about data technology and information technology as separate concepts. Context is all that's needed to understand what is meant by any of these terms.

None of this is to suggest that all information is equal, alike in value or usefulness—quite the opposite. Information comes in all shapes, sizes, and flavors. Some is great, some is terrible, and a lot is in between. Furthermore, some information is transient, of passing worth or significance, while some is more durable. Some is, at least to a commonly received notion, proven while some is mere conjecture, assumption, hypothesis.

Consider also that information does not depend on our activity or awareness. It stands alone with or without people in the picture. The precise population of a country exists even though we don't know it, as do the locations and amounts of oil reserves not yet discovered, the exact number of computer viruses detected on a given day, and the number of illegal downloads that would have resulted in actual purchases if illegal downloading was impossible.

Of course, untapped information serves no purpose. That's where people come in. It is people and their machines that seek, collect, create, and manipulate information to perform one or another function, and then presumably put the results into use.

So we come to knowledge

There is a profound difference between information and meaning. [11]

> **The brain and the mind**
>
> Is there any difference between the brain and the mind? Common usage says that the brain is an organ while the mind is the consciousness that derives from the processes of the brain.
>
> Thinking, perception, memory, imagination, emotion, and the like—those are products of the mind that result from the workings of the brain. So when the brain is functioning, the mind comes along.

Now, what to do about knowledge? We can draw some parallels between information and knowledge, but knowledge is rather different. We've seen some definitions that call upon the word knowledge to aid in defining information. Rather than equating or confounding the two, we find that a reasonable and useful distinction can be made by bringing the mind into play. When we do, definitions of knowledge actually tend to be rather in agreement, the idea being that knowledge is the mind's grasping or making use of (our definition of) information.

Here are some definitions of knowledge.

Knowledge: "information in mind: general awareness or possession of information, facts, ideas, truths, or principles." [12]

No distinction is made among the items the mind operates on, which is consistent with our inclusive definition of information.

> Knowledge: "awareness, understanding, or information that has been obtained by experience or study, and that is either in a person's mind or possessed by people generally."[13]

There is a big difference between possessing information and understanding it. Awareness and understanding elevate information. Simply knowing a fact is not the same as grasping its meaning. That's where the mind comes into play.

> Knowledge: "the fact or condition of knowing something with familiarity gained through experience or association."[14]

Interposing the mind raises information to the level of knowledge.

Where does information come from?

Information, in the factual sense, is all around us in what we call its native state—existing whether we know about it or not. We gather it by researching, observing, listening, counting, and so on.

We view objects in the universe, seeking clues to the origin of the earth; measure and test ancient bones to look for early ancestral linkages; compare chromosomes and illnesses to find susceptibility to diseases; collate meteorological data to forecast the impact of an impending hurricane; perform experiments with particle colliders to look for clues to missing matter in the universe; crash rockets into meteors to seek signs of the early universe; plumb the ocean depths in hopes of finding new life forms; survey people to assess product marketability; poll population samples to discern voter preferences; and much more.

It is important to keep in mind that although the actual facts of these phenomena exist, as sophisticated as our experiments and analyses might be, the meanings of what we see, or more accurately, what we think we see, are still just the interpretations of the moment filtered through the limitations of our equipment and our own predispositions. We hope to do a reasonably credible job but never know with absolute certainty because interpretation and judgment are involved.

Of course, our searches are not in vain. The quest to know more pushes us to make new discoveries and to seek additional evidence to support our views or to disprove them, leading to stronger interpretations that are closer to true facts, or at least good enough to be useful.

In teasing out what we can find about native information, we also create information in a broader sense by our various and several manipulations and invention. This takes us beyond the native, as far as to the completely artificial. We watch, we read, we estimate, we reflect, we imagine, we internalize, we discuss, we debate, we opine, we muse, we meditate, we reformulate, we write, we speak, we record, we accumulate, we dispense.

The bottom line is we do not sit idly by. It's not so much "I think, therefore I am"[15] but rather I am, therefore I think. Consequently, we continuously look for and produce more information, these days collectively in mind-numbing amounts.

We know you—data collection   What is there about us that isn't recorded somewhere? Hardly anything, it seems. As we shop, search, surf, blog, post, download, friend, tweet, chat, email, use our cell phones and smartphones, information about us is being collected. This happens with our knowledge and without, with our consent and without, for ostensibly beneficial purposes and not.

Have you paid a bill? We know that. Have you compared bank loan rates? We know that. Have you signed up at a local fitness salon? We know that. Have you moved? We know that. Have you returned a purchase? We know that. Have you coordinated a meeting with some friends at a restaurant? We know that. Have you taken a flight? We know that. Have you used GPS? We know that too, and where you went as well.

That there is no escape is a major issue with many dimensions, among which are privacy, security, data ownership and control, theft, inaccuracies, mischaracterizations, unauthorized use, false identity, anonymity, circumscribed data, and ethics. Because there are so many dimensions to this issue, we look at particular aspects in several places—the "And so—" section of this chapter and in Chapter 4, "That is to say—free expression and privacy"; Chapter 5, "What's mine is whose?—intellectual property"; Chapter 6, "You, me and everyone else—alone together and vice versa"; Chapter 7, "Attacks, bit by bit—from nuisance to cyber warfare"; and Chapter 9, "We are what we do—ethics and responsibilities".

US politicos and the media wellspring—information sources too   In 1933 there were no computers, no cell phones, no Internet, no social networks. But there was one network that had strong national penetration, could provide real-time communication and, if used effectively, could trigger feelings of a personal friendly connection to its listeners. It was called broadcast radio.

In 1933 radio was in the midst of its golden era. It also was the first year of Franklin D. Roosevelt's first term as president of the United States[16] and the year he began a series of radio broadcasts he called fireside chats, a practice he carried out for all eleven years of his administration. The charismatic Roosevelt used radio skillfully and persuasively to deliver real-time folksy messages right into the homes of American audiences.

Actually the continuation of a practice he began in 1929 during his first year as governor of New York State, his radio broadcasts were quite different from the formal speeches that typified what may be called official presidential addresses. Notwithstanding their intimate humanizing character, they dealt with such serious matters as the state of the economy[17] and the carriage of World War II.

At a time when there were many hugely popular radio programs, Roosevelt's broadcasts consistently had more listeners than any other show. With an engaging style, he seemed to be talking directly to the listener. The chats gave great comfort and reassurance to vast numbers of Americans during very trying times and built enormous confidence in Roosevelt as an understanding and capable leader. He certainly knew how to use the most powerful IT of the time.

Though the impact of those chats was clear, the next two presidents, Truman and Eisenhower,[18] made little consistent use of radio. That happened to coincide with the declining popularity of radio and the growing reach of television, but neither of them had the dynamism and charisma of Roosevelt, which may have explained their reluctance to use either medium consistently.

By the end of the fifth decade of the twentieth century, television networks had developed very strong national penetration, similar to radio in its golden era. John F. Kennedy,[19] another magnetically appealing speaker, realized that he could capitalize on the impact potential of television in much the same way as Roosevelt did radio. Kennedy used TV to great advantage in his campaign for the presidency and during his presidency to reach out to the populace in trying times.

The drama of his October 1962 broadcast telling the nation that the Soviet Union was building nuclear missile sites in Cuba, of his naval blockade of Cuba to prevent more supplies from reaching the island, and of his demand that Russia dismantle the sites and recall its supply ships or face attack was an incredibly riveting broadcast.

It was picked up by newscasters in many other nations and so served to alert them of an impending confrontation that had the world on the brink of nuclear war. The impact of a powerful IT that got the message out was a significant factor in building pressure on the two countries' leaders to reach an agreement and dial down the tension. The so-called Cuban missile crisis was resolved after 13 very tense days, when the USSR turned its ships around and agreed to dismantle the sites in exchange for a US promise not to invade Cuba.

After Kennedy there was another lull in the use of broadcast media by US presidents. Johnson, Nixon, and Ford[20] did not take to the airwaves consistently. President Jimmy Carter,[21] next in line, tried to emulate Roosevelt's fireside chats. He appeared on TV wearing a casual cardigan sweater and actually having a fireplace in the background, but he didn't have the charisma of Kennedy or the dynamism of Roosevelt, so it wasn't very effective for him. He didn't continue the practice.

Then came Reagan, two Bushes, and Clinton.[22] In a throwback move, they returned to radio to deliver mostly weekly broadcasts in a vein similar to Roosevelt's, even though television was more popular. The Bushes seemed more comfortable speaking on radio than appearing on television and so, came across well on the older medium. The homey appeal of Reagan and Clinton worked well on radio and television and they used both media effectively to their great benefit.

President Obama,[23] another charismatic speaker who conveys confidence and seems to focus directly on the viewer, became the first US president to make consistent and substantial use of new media and the Internet. In addition to capitalizing on social networks to build a groundswell of followers and to raise a great deal of money for his campaign, he has taken the fireside chats tradition to new media by posting videos on YouTube along with podcasts, live streams, videos, photos, and blogs on the White House Web site.[24] Available online at any time, this makes accessibility global, taking reach to a new level.

Many members of every government level follow similar paths, putting out tremendous amounts of information. Local politicians increasingly use new media to get their messages out. Governors and mayors have taken to the airwaves and the Internet as well. Some have turned their radio broadcasts into listener call-in shows, taking a lead from that successful radio and TV format, a trend that has proven to be quite popular. A prime example is New York City's Mayor Bloomberg,[25] whose radio show has a large audience despite his rather matter-of-fact this-is-the-way-it-is manner.

In open societies, independent political bloggers, Facebookers, and tweeters—politicians themselves, those who write for them, and especially anyone with particular interests and viewpoints to express—readily add to the flood of what we come across by taking advantage of new media's ease of use and ready availability. So it's not only the politicos who use the media to their advantage. The vast number of pundits and pseudo pundits build audiences not just from what they say but by encouraging comments and even rebuttals from their followers.

The same practices are followed in closed societies but with a major difference. Old and new media are state run, state supervised, and state monitored. Their governments exercise strict control over information flow, availability, and content, so we know that activity is almost completely limited to government-sanctioned participants disseminating government-sanctioned material. Dissenting opinions are quickly quashed and dissenters dealt with harshly.

There is a lot more to this, including the fact that control of media and IT is not just a phenomenon in repressive countries. It's present to some extent in every society, though not anywhere close to the extremes found in authoritarian ones. An important issue, we go into it in some depth in Chapter 4.

Truth or consequences

The speed of communications is wondrous to behold. It is also true that speed can multiply the distribution of information that we know to be untrue.[26]

The tremendous flood of information available, ostensibly to truthfully inform us in one way or another, is not necessarily truthful or informative. What we are being fed by whom and the difficult issue of separating the valuable from the essentially valueless is a serious issue. Even in open societies, what we get is biased in ways that put the best light on the disseminators or the institutions they speak for, whatever their beliefs. That's not much of a surprise, but whether it means deliberate misinformation, a clever spin, or actual fact, it is an important consideration to keep in mind when assessing information validity and value.

Let's look at the data/information/fact/knowledge mélange a little more closely, beginning with an axiom:

All information is not equal, nor is the import or impact of every kind of information we might come across, whether useful, misleading, or false.

Misleading and false information is all around us, part of daily life, but not all has the same effect. Consider a flawed recipe for chicken dumplings, a worthless though famously touted diet plan, a testimonial-laden pitch for an unproven miracle cure, claims that global warming is a myth, a blogger who jumps on a story to be the first to break the news only to find later that the stated "facts" were wrong.

Well, a poor recipe may taste awful but is probably not going to be dangerous. A worthless diet plan may set the dieter back a bit physically, may cost money for no results, or may be harmful to health. Skipping proper medical treatment by chasing false hope can have dire consequences in the short or long run. Not taking decisive action to alleviate global warming has huge long-term consequences for all of us. As for the blogger, did the report say that the local high school team won games they lost, that a product was not recalled that was or vice versa, that the cause of an airplane crash was a bomb on-board planted by a terrorist when there was no bomb and maybe not even a crash?

> On January 13, 2012, the cruise ship *Costa Concordia*, owned by the Carnival Corp., deviated from its prescribed course and crashed into a rocky outcrop off the Tuscany coast of Italy near the tiny island of Giglio. It quickly took on water, listed far over on its starboard side, and had to be evacuated of its 4,200 passengers and crew.
>
> One early report said the captain altered the course to allow a crew member to wave to his family on Giglio. A later one said he did so to show off the new ship to people in his home town. Still others say it was due to inattention. Some reports said the course change was not authorized by the company, while others said it was.

Then there are denials and deliberately delayed information. News stories have recently brought to light these three examples:

- A sudden acceleration problem that resulted in thousands of accidents, several fatal, was claimed to be due to driver error by Toyota. After denying reports to the contrary, in 2009 the company recalled millions of cars for repairs. The recall was subsequently expanded, adding nearly 2,200,000 more in 2011. By then, over 14,000,000 cars had been recalled.
- In August 2011, Cargill Meat Solutions Corporation, a huge meat processor, recalled approximately 36,000,000 pounds of ground turkey because of potential salmonella contamination, which went unpublicized until a large number of consumers fell ill. Recalled product dates stretched back six months, to February. Questions were raised about appropriate food handling, testing for food-borne pathogens, and timely notification of problems.
- On April 20, 2010, a fire and explosion on the oil drilling rig Deepwater Horizon in the Gulf of Mexico resulted in the sinking of the rig and a massive gushing of crude oil from the well. BP, the global oil and gas company that licensed the rig from Transocean, laid the cause of the leak to a failed blowout preventer, but some rig workers said that the well was leaking well before the explosion. BP initially reported very low estimates of spill volume, though within two days a five-mile-long oil slick was apparent. Coast guard estimates

of spill rates escalated from 1,000 barrels a day to 5,000 by the end of April. In early May, the Gulf coast was affected. Spill rate estimates in late May by various experts testifying before a Congressional committee ranged from 20,000 to 100,000 barrels per day. By mid-July the leak was finally stopped. BP agreed to set aside $20 billion for damage claims.

Over the years, politicians have refused to admit to unethical or illegal activity until the evidence becomes too convincing to ignore. Here are two examples:

- On June 16, 2011, New York Democrat Anthony Weiner resigned his Congressional seat, finally admitting to long-term sexually oriented relationships with women via Twitter and Facebook. He had denied the evidence, claiming at one point that hackers had placed false information on his pages. After several months of mounting evidence, he relented, acknowledging the truth of his dalliances.
- Mark Sanford, former Republican governor of South Carolina, admitted to an extramarital affair after he was called out on his six-day disappearance in June 2009. The official statement on his absence was that he was "hiking the Appalachian Trail." It turned out he was in Argentina visiting the object of his affections. The "hiking" phrase became a euphemism for having an illicit sexual relationship.

Sexual improprieties aside, many politicians have taken advantage of their positions to secure campaign donations, realize gains from insider trading information, receive gifts in exchange for favoring views of lobbyists, and a host of other practices bordering on or exceeding the boundaries of malfeasance. Two examples:

- In 2010, after nearly two years of refuting charges of ethics and tax violations, Representative Charles Rangel, Democrat of New York, was censured by the House of Representatives.
- In 1997, after an embattled but losing defense of an accusation of 84 ethics violations, Newt Gingrich, Republican congressman of Georgia and speaker of the House was sanctioned by the House ethics committee decision that included a $300,000 penalty. He attributed the outcome to bad lawyering and overzealous ethics committee, although half of its members were Republicans.

What we are faced with in the information deluge is more than separating the wheat from the chaff. We'd like to be able to recognize the so-so, the good, and the great wheat as well—that is, to properly evaluate the information we discover. While that always was the case, the problem is multiplied enormously by the Internet, where any of us with a connection can be creators or propagators of fact and fiction. Information from just one person, whether accurate or misleading, true or unintentionally or purposely false, can rocket around the world in an instant, repeated by a great many others. This is unprecedented in the history of communication.

Fact, fiction, and the in-between—evaluating information

There are no gatekeepers to filter truth from fiction, genuine content from advertising, legitimate information from errors or outright deceit. . . . When we are all authors, and some of us are writing fiction, whom can we trust?[27]

Quod gratis asseritur, gratis negatur.[28]

> ### Theory: a guess, a fact, or ?
>
> In ordinary usage, theory is synonymous with speculation, hypothesis, possibility. "My theory is that the color yellow makes people feel happier" or "My theory is that when a cat's tail is bushy it's going to rain."
>
> To scientists and philosophers of science, a theory is that which has undergone rigorous testing by experts in the field and cannot, so far, be falsified. There is always room left for the discovery that refutes the theory, but as successful replications pile up, it finds firmer and firmer ground. So, for example, the theory of gravity is not a guess, not a supposition, but as close as we can get to a proven fact.

How are we to judge the value, the worthiness, of information? Is it valid, truthful, accurate, falsified, misrepresented, illegitimate, relevant, complete, current? Timeliness is an important aspect as well. Some information is time insensitive—physical constants, mathematical relationships, the SAR rating of your cell phone, census data, historical documents, survey data, results of experiments—although opinions and conclusions may change when data are re-examined.

Other information loses value quickly as it ages—a weather forecast, today's TV program schedule, job opening lists, current mortgage rates, funding opportunities, stock prices, the traffic jam on your commuting route, the story your friend is about to tell you for the fifth time—though some of those can yield useful insights on further scrutiny.

Knowing the source counts. Some Web pages at least show the author and date of the page. Many do not provide even that minimal information, let alone cross-references to corroborating material, links to sources and related information. Other sites make a point of telling us those things, or at least a good number of them.

But even when there are specifics, how are we to assess their legitimacy? Perhaps we could hark back to the pre-Internet days for some guidance, when print publications dominated as primary sources. Seeing something in print—a book, a newspaper, a journal—seemed to confer legitimacy. Though that conclusion was not always justified, can any of the ways we ascertained the authenticity, the validity, of what we read in print publications apply to information we read online?

One criterion frequently noted is, does it come from a reliable source—for example, peer reviewed academic journals, longstanding highly regarded publishing houses, reputable newspapers? That seems to be a good idea, but haven't we learned of falsified data, poorly checked reportage, and even outright hoaxes coming from just such sources?[29] Still, they are the rare cases, so it is an idea that adapts well to online sources and has more merit than simply taking it on faith.

Another idea is that if there is more than one source with the same information we can be more sure of its reliability. We know that it's far easier to search for such information conformity on the Web than it is among print sources. But wait a minute—many Web sites get their information from other Web sites and back and forth in a round robin. So a piece that is of questionable soundness appearing in one source can very quickly become replicated in whole or part on a great many other sites, hardly an indication of validity.[30]

So many questions, so few answers. One thing to count on—we can always be fooled. Here are some reasons why, some related areas affecting our analyses of and confidence in information.

We are what we believe—selective narrowing

> You can't convince a believer of anything; for their belief is not based on evidence, it's based on a deep seated need to believe.[31]

> Faith is the antithesis of proof.[32]

We all have biases. This is not automatically a bad thing. Bias simply means preference, opinion, viewpoint, preconceived notions about the way we think. Even the most upstanding, ethical, and moral of us have biases. In fact, believing that ethical behavior is proper is itself a bias.

Many biases are innocuous—I prefer the color blue; I like broiled fish more than fried fish; I would rather live in the city than in the suburbs; I root for my team and against yours. A good number of these partialities operate at an unconscious level— I am likely to buy the blue shirt rather than the green one just because I find it more appealing, not because I ponder the essence of blue; I usually will broil fish rather than fry it because it's more tempting to me, not because I've done some impartial analysis. For the most part, we don't think much about these kinds of choices. Such inclinations of ours are socially dispassionate.

Some biases, of course, are socially and personally damaging—racial, age-related, religious, and lifestyle prejudices for example—because they are characterized by unfairness, unequal treatment, even demonization. These are particularly tenacious, as are any for which belief overrides facts. Discussion becomes irrelevant. A great many upheavals in societies can be traced to prejudicial biases of these sorts.

For relatively objective quests—the closing Dow Jones average last Friday; the exchange rate between the dollar and the euro on a particular date and time; the county that had the highest average rainfall in 2011; the state that currently has the lowest property tax rates—opinion has little influence, though even with these relatively straightforward items there can be definitional and measurement issues.

Brain bias[33]—a complication   When it comes to information about which we have an opinion, especially when that opinion is based on strongly held attitudes— Which political party has better ideas? Should senior citizens pay school taxes? What role should government play in public health policy? Should libraries censor Internet use? Where is the best place to live? Who's investment advice should we follow?—we are predisposed to put more credence in sources that support our opinions rather than those that run counter to them, regardless of source.

As a matter of course, we are more apt to seek information that conforms with our biases, to read it more completely, to believe it more readily, and to internalize it more fully. Truth, like beauty, often is in the eye of the beholder.

By the same token, we are more likely to avoid, or at least give short shrift to, non-conformant information —that which is not consonant with our biases. Such is often the force of cognitive dissonance. By selective attention, focusing on what's consonant with our beliefs and purposely avoiding what isn't, we can avoid potential dissonance, or at least reduce it after the fact. Presumably we also can reduce or eliminate dissonance by suspending our beliefs, but that's much harder to do.

> *Cognitive dissonance* is feelings of tension, anxiety, or discomfort that come from mentally confronting conflicting, clashing thoughts, ideas, or information. It's commonly felt after making an expensive purchase when we wonder if we've spent our money wisely. Buyer's remorse is a popular synonym for post-purchase cognitive dissonance.
>
> The same concept applies to more critical choices—who we vote for, our spousal preferences, the medical treatment to pursue, investment decisions.

Can we suspend our beliefs?   Our biases have consequences for our choices and decision making. Depending on the particular circumstance, that can be important or not. When election campaigns are underway, do we listen only to the speeches of candidates in our favored party and treat them as truthful and proper, the right stuff? If we do listen to opponents' speeches, do we dismiss or deride what they say? How about a debate between our candidate and theirs—do we assess either of them objectively? Can we?

Much of what our brains process is filtered through our biases, and much of that filtering is unconscious, or at least subconscious. While we can make a determined effort to confront our prejudices, we can't operate in any sphere in a completely objective way.

We know that some arguments are persuasive enough to change our minds, especially when presented by adroit writers, or even more effectively, by charismatic speakers. But are we being responsive to the arguments themselves or our favorable bias towards the arguer?

Suppose we convince ourselves that we need to hear what the opposition has

> Ptolemy (c.85–c.165) observing the transit of the sun, concluded that the sun revolved around the earth—a view concordant with religious doctrine that the earth was the center of the universe. Galileo (1564–1642), observing the same phenomenon, concluded that the earth revolved around the sun, for which he was accused of heresy. Two ostensibly unbiased observations led to opposite results. Even so-called objectivity often needs proof.

to say. Even though we can't listen completely objectively, we can use what we hear to formulate counter arguments that conform to our original beliefs and diminish the value of ideas in opposition. That's a version of the concept behind "know thine enemy."[34]

What this all boils down to is a simple question: Is there such a thing as true objectivity? In some realms, especially mathematics and some empirical sciences, we can reasonably separate the subjective from the objective. So we can say without bias that 1 + 1 = 2 (leaving aside esoteric discussions of what numbers mean), that gravity exists, that the earth rotates on its axis and revolves around the sun.

Once we move to opinions—that the electoral college is the best way to conclude an election, that small government is better than big government, that zoos are a form of animal cruelty—we're on less firm ground.

Understanding numbers—reading between the lines

Not everything that counts can be measured. Not everything that can be measured counts.[35]

To this we add two corollaries:

Not every measurement is what it claims to be.

Not every measurement measures what it appears to.

There are many things we'd like to measure but can't. It is too costly and time consuming to measure everything that we can, but much of what we do measure is of questionable worth. There's a big difference between measuring for measurement's sake and measuring to produce information of value.

There is a huge amount of measurement-based information zipping around the Internet. We know that some is worthwhile and some is worthless. What can we say about that? Let's talk about numbers, the paragons of facts—maybe.

An aura of accuracy   We are predisposed to view numbers as providing precision, truth, and validity to an accompanying statement. Without numbers, assertions can seem to be opinions, whether or not they are factual. With numbers, they appear to be straightforward and precise. If you want people to believe something, whether it's true or not, put a number on it. If you want to add credibility to a statement, whether it's credible or not, put a number on it.

How is it possible that every auto insurance company can *save you $387 or more* compared to the others? How much more compelling is the magazine that flaunts *264 ways to save time around the home* on its cover in contrast to the one that simply says *how to save time around the home*. Such precise-looking numbers grab attention and must be true.

Then there is the unlimited ends statement:

● loans of up to $20,000 or more!

That actually means loan amounts ranging from $0 to infinity. It would be just as accurate to say *loans of up to $1 or more,* or *loans in any amount.* Of course, none of those is what the lender means. There are lower and upper limits on what is really offered, but by popping in what the advertiser believes is an eye-catching number the come-on, while still meaningless, becomes stronger.

Impact from imprecision is another strategy. Compare these statements:

- Cigarette smoking is strongly related to death from lung cancer.
- This year 400,000 people in the US will die from lung cancer due to cigarette smoking.

The first statement, though not specific, is considered by researchers to be factual, based on years of investigation, but since it's numberless it seems like an opinion that can be dismissed as such. The nice round number in the second statement is an estimate. Yet 400,000 makes the assertion specific, accurate looking, and more impactful, whether it's close to the truth or not.

How about these:

- A majority of Americans are overweight or obese.
- More than 34 percent of Americans are obese, 32.7 percent are overweight, and just under 6 percent are extremely obese.[36]

Again, the first is a general statement considered to be factually based, but the lack of numbers makes it seem like it's just a supposition. Precision is implied in the second, especially the number with the decimal point, but how precise is it? On the other hand, does it matter? It depends on intentions—to clarify, to grab attention, to mislead.

When considering numerical statements, it's useful to know how the numbers came about. For the last example, let's see how overweight and obesity are measured. According to the Centers for Disease Control and Prevention:

> Overweight and obesity are both labels for ranges of weight that are greater than what is generally considered healthy for a given height.[37]

This moves us from precise to "generally considered," whatever that means. So here's a precise definition:

> [The Body Mass Index (BMI) is a] measure of body fat that is the ratio of the weight of the body in kilograms to the square of its height in meters (a [BMI] in adults of 25 to 29.9 is considered an indication of overweight, and 30 or more an indication of obesity).[38]

Is that mathematically precise BMI universally applicable? No, because the assumption is that it is an accurate measure of body fat ratios. For some people it is, for others it isn't. And what about the precision implied? If your BMI is 24.9 you're OK, but just .1 more and you're overweight.

Of course, definitions at boundaries often are a problem, so instead of a literal interpretation, think that if you're approaching a boundary, do something about it.

> It is widely held as true that there is a serious weight problem in the US and other parts of the world. We don't dispute that, nor do we mean to devalue BMI. While it isn't universally applicable, it's quite useful and until something better comes along, reasonably functional, at least as a starting point.

Precision may not be relevant   In many instances, precision is not the goal. Numbers are used simply to provide reference points. Consider:

> The normal temperature for today is 48 degrees.

Normal is just an average of the day's high over many years. Does it matter if it's precise? After all, a long-term average is not affected by recent trends, which may be more significant to know. But its purpose is to give us some idea about today, a yardstick to compare today's reading. That's a point of interest that's usually enough.

How about:

> The typical family spends 13 percent of their disposable income on food.

What's a typical family?[39] Does it matter if spending is a little more or a little less? No, because the statement is meant to give us a sense of how much of our income we spend to feed a family. Of course, what we actually spend depends on our particular circumstances, but the number doesn't need to be very precise. It's doubtful that it could be, but unless it's wildly off-base, it provides a useful point of reference.

Extreme values render averages misleading   Ponder this:

> When Bill Gates walks into the room, we're all instant millionaires, on average.

That's the force of extreme values. To make sense of the statement, we need to know how variable the underlying population is—the range of the data, whether the values are relatively evenly distributed or severely skewed, whether there are one or more clusters, and so on. Statistical techniques take these and other factors into account, but by seeing only the average we don't know the whole story.[40] Too often, that's all we get.

Quite often extreme values are considered to be outliers—abnormal data points that should be considered atypical and therefore not relevant. Sometimes that's just what they are. After all, when considering the average net worth of a group of people, how likely are we to be including a multibillionaire? And if we are, isn't it more informative for our general purposes to leave that person out?

But extreme values may be more than anomalies. There is the possibility that they are the most important data points of all, illustrating unexpected situations that need to be taken into account. Key questions: Do those values contain information pointing to surprising conditions that warrant close attention? Are they irrelevant to our study and therefore safe to ignore? Was there an error in measurement, interpretation, recording that resulted in those atypical values? Were there forecasting assumptions that didn't hold?

Forecasts presume continuance   There are many forecasting techniques that have a good track record, based on performance measured by predictions that turned out to be reasonably correct. So, they are depended on as being fairly reliable. They all presume some sort of observable stable pattern.

Prediction is based on seeing how past events indicate forthcoming events, from which we also derive how a current event is a precursor to a future event. All assume that precedents hold. When they don't the forecast fails. Perhaps we didn't see the actual pattern, missed a shift, had bad information—or perhaps it was that key extreme value that we ignored.

It's not hard to imagine situations where the shift is more important than a continued pattern, when a seemingly stable pattern changes. Suppose we use a naive method of weather forecasting: tomorrow's weather will be the same as today's. Measured by percent of correct forecasts, we'll do pretty well, in the neighborhood of 80 to 85 percent accuracy. But we'll have missed every time the weather changes, which often is the most important thing to know.

The 100-year flood wasn't due for another 70 years, but we were inundated yesterday. Chernobyl was operating as expected, and yet, meltdown. The housing bubble won't burst yet, and then it did. Blowout protectors for undersea oil drilling platforms in the Gulf of Mexico will prevent huge oil spills, until they didn't. Forty-foot high seawalls will protect Japan against projected tsunamis, and then along came a 60-foot tsunami.

Correlation is not causation   It is well known that correlation merely indicates that one factor and another usually appear together. Still, it is easy to make a mental jump to causation, so it bears noting that correlation does not demonstrate that one event caused another. Here's an example:

> Last year 34,000 people who were bullied via the Internet killed themselves.

The statement implies a link was found between bullying and suicide, that when the one occurred, so did the other. It seems to follow, then, that bullying causes suicide. Setting aside the complication of how bullying, not an all or nothing proposition, was defined, suicide is rarely a single-cause event. While it was discovered that these 34,000 people were bullied and also committed suicide, we don't know that bullying was the precipitating cause. It's likely that it was the principal cause in some instances, an exacerbating factor in others, and not at all the reason in still others. Just based on the statement, we don't know, but whatever the case, we can't assign cause.

Comparability counts   Suppose we read a report about a finding that the percentage of students who scored above state averages in reading and math was higher in one high school than another. The conclusion drawn is that teachers in the school with the greater percentage are doing a better job of educating students.

That may be true if the student bodies of the two schools and the schools themselves are comparable on a number of dimensions, including the student composition and circumstances, class sizes, support facilities, environment, home situations, teacher backgrounds and experience, curriculum equivalence, and so on. If comparability was assumed but not investigated, the comparison could be misleading and the conclusion unwarranted. And if left unstated, as it so often is in the reports we see, how are we to judge?

Missing information   It can happen that a true statement is still a misleading one because the whole story is not presented. Consider:

> Since the No Child Left Behind Act of 2001, average reading scores have improved each year.

That being true, we conclude that the Act is successful. But what's not stated is that average reading scores also improved for some years before No Child Left Behind.

> By 2040 Manhattan will be under water due to global warming melting the polar ice caps.

Calculations reveal that this could be true, but not noted is that it requires global warming sufficient to melt all the polar ice. Will that happen? Is it an extreme outlier we should ignore or a possibility that warrants careful attention? Would it be a relief if only some of Manhattan winds up under water?

At the same time, be wary of misleading comparisons due to incomplete information.

> In a race, our runner finished second but his main opponent came in next to last.

Left out: it was a two-man race.

And how about real estate ads that tout *only 30 minutes to New York!* Does that mean town center to mid-Manhattan, or border to border, or ? And is it during rush hour, midday, late at night? Does the clock start when you leave home or when the train leaves the station and does it stop when the train arrives, when you get off, or when you reach your workplace? What about a bus? What about a car?

Rounding does not mean loss of accuracy   Mathematical calculations involving division or multiplication by fractions can result in many places after the decimal point. A number like 51,341.62436143 looks super accurate, but in most cases it's a false accuracy. All those decimal places are not necessarily meaningful, but merely the result of arithmetic.

A general guiding principal is that calculations don't produce more accuracy than is present in the original numbers, whose values in turn depend on the accuracy of our measurements. So if our data are whole numbers, round to the nearest unit. If they are accurate to two decimal places, round to the second place.

Except for special circumstances, rounding is considered to be sufficient. How many places to carry is situation dependent. Except for a few specific calculations, the Internal Revenue Service (IRS) is satisfied with rounding to whole dollar amounts on income tax forms. Bank statements typically are rounded to the nearest penny, even though interest calculations can generate many more decimal places. What do you suppose happens to those fractions of a penny that we don't see in our balances?

Midpoint values are usually included in rounding up. So 12.5 would be rounded to 13. Values less than midpoints are dropped, or as some would say, rounded down. So 12.4 becomes 12. This can produce what are called rounding errors, which occur when rounding in the same direction in chains of calculations cumulate.

Some accounting practices call for carrying many decimal places until the final result to minimize that sort of error. Computers using floating point arithmetic[41] can produce rounding errors as artifacts of the process. But neither of these situations is significant in most instances.

Rounding does not only apply to decimals. We may round to the nearest nickel because we don't want to deal with pennies. If we're calculating average commuting time, we might round to the nearest minute. Even that might imply greater than warranted accuracy. When asked how long commuting takes, we may get answers like about 30 minutes, or an hour or so. A few minutes off is not critical. It is, however, meaningless to measure commuting time as 27 minutes and 38 seconds.

The point is, don't be fooled by appearances. If you want to know how accurate a value is, find out how accurate the numbers are that went into the calculation and how the calculation was carried out.

Context eliminates vagueness When we use superlatives, like better, smaller, greater, wiser, a comparison is implied. Without supplying the comparison, the expression is vague.

> I feel better now.

Better than when I had the flu? Than when I mistakenly thought I was being fired? Than I did before my team clinched a playoff spot?

The same applies to statements without context.

> That fellow is quite short.

He's 6′3″ tall but we're talking about small forwards in the National Basketball Association (NBA), where a height of 6′9″ is not unusual for the position.

> He's not heavy enough.

He weights 294 pounds but he's a defensive lineman in the National Football League (NFL) where 325 pounds is not uncommon.

How many products have you seen that say *new and improved* on the package, but don't tell you what was changed? Sometimes it's only the packaging. Similarly, *stronger tissues* implies a comparison to competitor products, but may only be compared to the company's own prior version. Vagueness is a common tactic in advertising. The most clever ads make vague statements look definitive.

And so—we are what we know

information overload.

Everybody gets so much information all day long that they lose their common sense.[42]

It's not just that IT has its benefits and drawbacks, it's that they're often two sides of the same coin. What we celebrate on one hand, we rue on the other. Compared with all of history, we are in a golden age of information. We have fingertip access to more data, records, statistics, images, video, audio, opinions, and reportage than was dreamed of only a few decades ago. Yet that same access has created a bind—we cannot comprehend the scope of it all, let alone validity.

Information overload is not an idle concept. Since too little information leads to bad decisions, we may want to believe that the more we see the better our decisions will be. Some studies have shown that this is not necessarily so.[43] While the benefits of our marvelous access to information are indisputable, as we pile on more and more there are sharply diminishing returns. Going beyond a saturation level of information or number of choices quickly reduces the goodness of our decisions—we actually do better when there is less confronting us, even with incomplete information, than when we are overloaded.

Such decision-making difficulties have ramifications for businesses as well as for individuals. When consumers are bombarded with come-ons for similar products, the result may be postponing a purchase decision or declining to purchase at all. The vast arrays of equivalent products in supermarket aisles may actually reduce sales, especially for purchases of non-necessities. In an interesting choice experiment, shoppers facing 24 different jams were only 10 percent as likely to buy any than those who had only six options.[44]

Despite that phenomenon, marketers wage vigorous battles for supermarket shelf space. The more varieties displayed, the more space, so more is better is the name of the game. Where there once was white, wheat, rye, and pumpernickel bread, now there is a dizzying array of types and brands that easily take up an entire aisle in a large supermarket. The same can be said of toothpaste, soup, cereal, and even pasta, as well as many other products.

It's not just supermarkets. A barrage of choices confronts us in nearly every type of physical and online shop, when we're at work and when we're at home. Watch a TV commercial for a prescription drug or read about one in a magazine or Web ad and you will find them accompanied by a long list of possible side effects that can make the cure seem worse than the ailment. And it's not just ads—you get an even more extensive list when you fill a prescription.

Of course we need information to evaluate risks. Forewarned is forearmed, right? But is information helpful when it's contradictory? Here are some common side effects for a single medication: *constipation, diarrhea, drowsiness, trouble sleeping*. What are we to make of that? Then there's the coverall *do not take if you are allergic to any of its ingredients*, as though we understand what the obscurely named ingredients are. And what does the euphemistic *may cause a fatal episode* do for you?

The sheer number of side effects noted is another issue. Two examples[45]: Zoloft, an antidepressant, has 17 common and 45 less common side effects; Aleve, a pain treatment, has 10 common and 46 less common. For both, "This is not a complete list . . ." Some appear to be counterintuitive—for example, Aleve notes *unusual joint or muscle pain*, which seems odd for a joint pain treatment. These examples are not meant to single out Zoloft, Aleve, or any other drug. The lists are typical. They are so extensive because they include rare and extremely rare effects. The question is, what's the point of all this not very helpful information? The Food and

Drug Administration and congressional mandates explain some of it. Could it also be pharmaceutical companies protecting themselves against lawsuits? Think about it.[46]

Consequences   There's more to information overload than postponed or poor decisions. The effects of having too much information are exacerbated by increased stress from having to sort and choose, by mental fatigue, and other related factors. Stress has physical as well as mental consequences. So we must apply our own filters to reduce information intake and choice to manageable levels. That way we can avoid those diminishing returns and the extreme of brain freeze, a kind of mental paralysis where we are pushed beyond making bad decisions to being afraid or unable to make any decision.

So filtering is imperative—and there's the rub. When resolving problems, finding our way through a deluge of information raises a difficult question: How much of what kind of information do we need? We want to make well-informed decisions based on appropriate information—valid, timely, and complete, of sufficient quantity but please, not too much.

To make the best choices we need to stay below our overload threshold, but that's not the whole story. What matters greatly is how our filters work. Common sense tells us we should disregard the least relevant and most questionable. Having the right amount of the wrong information won't do. We're better off with an incomplete but pertinent array of valid data.

So we hope for a reasonable number of the most valuable options supported by a reasonable amount of information. While it's not always clear what "reasonable" means, with experience, we come to know what reasonable is for us. On the other hand, our filters can't be perfect. We know that distinguishing the most applicable from the less relevant and the valid from the invalid is no mean feat. Our built-in biases tend to let pass information that's consonant with our beliefs and inclinations and to block or diminish the impact of information that's not.

Whether conformant information is incorrect or misleading can escape the purview of our filters. In the end, what we know may be as much a product of spurious information as anything else.

The business difference—when too much information isn't   Are the deleterious effects of information overload on decision making overstated? Erik Brynjolfsson, Lorin Hitt, and Heekyung Kim think so when it comes to businesses. They consider a deluge of information to be an opportunity, not a problem, at least for smart business managers. In their recent research they discovered that performance is higher in "firms that emphasize decisionmaking (*sic*) based on data and business analytics."[47]

Furthermore,

> Using detailed survey data on the business practices and information technology investments of 179 large publicly traded firms, [they found] that firms that adopt [data-driven decision making] have output and productivity that is 5–6% higher than what would be expected given their other investments and information technology usage.[48]

Their study compared firms that primarily rely on analysis of volumes of data to those whose managers depend heavily on experience and insight. The lesson derived from their work is that while businesses run on information, they thrive only if it's the right information and it's used properly. A major key is powerful data mining and analysis techniques.

Personalization—the answer or the question?   Companies are happy to help us keep our heads above the information flood. Knowing so much about what we do, what we like, and what we don't like makes it easy. Just feed that information into specialized algorithms and out pops a personal profile of each of us. Then when we browse for a movie, song, or book, when we search on some term, when we seek information about a news event, when we compare electronic devices, when we look for a restaurant, a show, an airplane flight, or just about anything, the first choices we see are likely to be consistent with our preferences.

The more we surf, search, purchase, and return, the more we rate, comment, declare likes and dislikes, praise and complain, the more focused our profiles become. Consequently we see more closely directed ads and more focused choices. We have less of a flood to wade through and come across fewer pieces of dissonant information. So that's the good news, right? Well yes, but as always, there also is bad news. One is that whatever we do is assumed to mirror us, yet that's not always the case.

If I purchase a gift on a Web site, something that I have no personal interest in and may even dislike, then that becomes part of my profile. I will get hints, suggestions, emails, tweets and the like aimed at that category of items. Some of the flood leaks through—alright, not so serious, but how about this next point.

We don't know what we're missing, that idea, that piece of information that could excite us, give us a new perspective, move us in a new direction. But it's atypical for us so we don't get to see it. It's not just ads and purchase suggestions, but information of all sorts—different views on the news, current events, politics, science, art, the world economies, and the like. Are we better off warm and cozy in our blanket of consonant information, or if we are challenged by confronting our viewpoints?

What do you think of this? *Vast Wasteland redux*

In 1961, Federal Communications Commission (FCC) chairman Newton Minow delivered an address that came to be known as the Vast Wasteland speech, from a line he used to describe the state of TV programming. Aside from the many fluff shows and violent content that typified TV broadcasting, he decried news programs for their focus on fires, crashes, and crime reports to the exclusion of the weighty and important issues of the time.

Minow ardently supported federally subsidized public service programs—commercial-free delivery of all sorts of information an informed populace should have—dedicated to just such issues, treated in depth by knowledgeable and experienced people. Importantly, his call was for independent news sources and analysis, not mandated government-produced programming or profit-motivated commercial broadcasting.

Minow realized that pressing the cause wasn't enough to make it a reality. He wanted to apply FCC's broadcast licensing authority over airwave use to require a certain amount of public service programming as a condition of licensing. Six years after his speech, Congress created the Corporation for Public Broadcasting, a private non-profit entity to oversee federal support of public television and radio. Numerous stations and programs were spawned as a result.

It is interesting to note that many in Congress are now opposed to supporting public broadcasting and are maneuvering to reduce or eliminate much of that funding. Their arguments often include the assertion that such broadcasting is biased in favor of those with viewpoints different from theirs, a claim that has yet to be verified and one more example of avoidance of dissonant information.

Attention everyone—now hear this If we're not aware of something, it has no impact on our views or decisions. But the more we try to pay attention to, the less time we have to give to any of it. We divide our focus into thinner and thinner slices, with the result that little gets sufficient concentration and certainly no sustained contemplation. What is not in current view recedes in awareness. That's why we can text as we walk along a street and bump into a lamppost, or why we can talk on a cell phone while driving and wander across the roadway lanes.

We like to think of ourselves as multitaskers, but for anything that requires the same parts of our brains to accomplish, we are at best sequential taskers. We can rapidly shift from one activity to another but in doing so we devote full or most likely only near-full attention to just one task at a time, and for a limited time at that. The higher the stakes for any particular activity, the more damaging is the lack of unbroken considered attentiveness.

An outgrowth of the divided attention phenomenon is the popular sentiment that the Internet is the prime culprit, that its growing use has tremendously reduced our patience as it has similarly reduced our attention span. The real issue is not, as many believe, whether the Internet has created some sort of attention disorder, some mental deficit. Rather, it's whether we've gotten into bad habits.

We have little forbearance waiting for query results; we have diminished tolerance for idle time; we must constantly receive status updates of all sorts, trivial or otherwise; we cannot be offline lest we fail to notice something or be unable to immediately update our own feeds; we keep our mobile devices turned on and at hand 24/7 so as not to miss or even delay getting the latest "intelligence" no matter how inconsequential, no matter how postponable; we have access to more information than we can possibly process but nevertheless we must have it now.

Who's on it first? Feeding our appetite for instant information puts tremendous pressure on news producers to generate reports without delay. One result is that all news becomes breaking news, to be posted immediately, leaving no time for fact checking, contemplation, analysis, or completeness. Misreporting is common even if unintentional. Once it's out, it never disappears, but is often picked up and repeated by other sources. So when we search, we are as likely to find the erroneous as not.

Beyond news reporters, all manner of us are unvetted content producers—blogging, podcasting, proffering opinions and reviews, creating video and audio and the like. Hardly any of these offerings are independently edited or verified. Often producers and sources are not identified. Deliberately or inadvertently, content may be mistaken, falsified, distorted, or misleading. Since misinformation spreads as easily via the Internet as the factual, we're creating a wildly inconsistent mass of data. Any perusal is as likely to turn up the not so good as the good. So in the end, *caveat lector.*[49]

Notes

1   Richard Dawkins, as quoted by James Gleick in "What Defines a Meme?" *Smithsonian Magazine*, May 2011.

2   William Pollard (1911–1989), professor of physics, University of Tennessee, and Quaker minister: http://thinkexist. com/quotation/information_is_a_source_of_learning-but_unless_it/226524.html

3   http://www.bing.com/Dictionary/search?q=define+data&FORM=DTPDIA

4   http://searchdatamanagement.techtarget.com/sDefinition/0,,sid91_gci211894,00.html

5   http://webopedia.com/TERM/D/data.html

6   http://dictionary.cambridge.org/define.asp?key=data*1+0&dict=A

7   http://www.bing.com/Dictionary/search?q=define+information&FORM=DTPDIA

8   http://searchsqlserver.techtarget.com/sDefinition/0,,sid87_gci212343,00.html

9   http://dictionary.cambridge.org/define.asp?key=40689&dict=CALD&topic=information-and-messages

10 http://www.merriam-webster.com/dictionary/information

11 Warren G. Bennis (1925–), psychologist, business leadership expert: http://www.brainyquote.com/quotes/quotes/ w/warrengbe133207.html

12 http://www.bing.com/Dictionary/search?q=define+knowledge&FORM=DTPDIA

13 http://dictionary.cambridge.org/define.asp?key=knowledge*1+0&dict=A

14 http://www.merriam-webster.com/dictionary/knowledge

15 René Descartes (1596–1650), philosopher, in *Discourse on Method,* 1637.

16 Franklin Delano Roosevelt (1882–1945), 1933–1945 term of presidency—the only three-term president in US history. The 22nd amendment to the Constitution, ratified in 1951, limits the presidency to two terms.

17 The US was in the midst of a great depression that lasted from the stock market crash of 1929 until about 1939 or 1940, when World War II kicked in and greatly changed population and workforce profiles.

18 Harry S. Truman (1884–1972), 1945–1953 term of presidency; Dwight D. Eisenhower (1890–1969), 1953–1961 term.

19 John F. Kennedy (1917–1963), 1961–1963 term of presidency, cut short by his assassination.

20 Lyndon B. Johnson (1908–1973), 1963–1969 term of presidency; Richard M. Nixon (1913–1994), 1969–1974 term; Gerald R. Ford (1913–2006), 1974–1977 term.

21 James E. (Jimmy) Carter (1924–), 1977–1981 term of presidency.

22 Ronald W. Reagan (1911–2004), 1981–1989 term of presidency; George H. W. Bush (1924–), 1989–1993 term; William J. (Bill) Clinton (1946–), 1993–2001 term; and George W. Bush (1946–), 2001–2009 term.

23 Barack H. Obama (1961–), 2009– term of presidency.

24 www.whitehouse.gov

25 Michael R. Bloomberg (1942–), 2001–2013 term as the 108th mayor of New York City, its only three-term mayor.

26 Edward R. Murrow (1908–1965), broadcast journalist: http://www.brainyquote.com/quotes/quotes/e/edwardrmu 116312.html

27 Andrew Keen, The Cult of the Amateur: How blogs, MySpace, YouTube, and the rest of today's user-generated media are destroying our economy, our culture, and our values, Crown Business, 2008, p. 65.

28 *What is freely asserted is freely negated.* Or as Christopher Hitchens (1949–2011), journalist and author, put it, "What can be asserted without evidence can be dismissed without evidence."

29 For accounts of retractions and a variety of dishonesty in articles printed in scientific journals, see the blog: http://retractionwatch.wordpress.com/

30 Some interesting ideas for evaluating Web pages are discussed at: http://www.lib.campbell.edu/collections/ webevaluation.html; http://plagiarism.umf.maine.edu/valid.html; http://lifehacker.com/software/feature/seek-and-ye-shall-find-how-to-evaluate-sources-on-the-web-137843.php

31 Carl Sagan (1934–1996): http://thinkexist.com/quotation/you_can-t_convince_a_believer_of_anything-for/191321.html

32 Edward J. Greenfield, justice (retired) of the Supreme Court of New York State.

33 Other terms are conformation bias, confirmatory bias, cognitive bias, and differential attention bias. Some consider these to be synonyms; others feel that they are different aspects of bias phenomena. For an interesting summary of a wide variety of biases, see: http://www.brainbiases.com/

34 Not a literal quote, but a sentiment espoused by Sun Tzu (c.544 BC–c.426 BC), Chinese general and tactition, in *The Art of War*.

35 Attributed to Albert Einstein (1879–1955), physicist and recognized genius.

36 Based on numbers from National Center for Health Statistics 2005–2006 data: http://www.reuters.com/article/2009/01/09/us-obesity-usa-idUSTRE50863H20090109

37 See the Centers for Disease Control, Body Mass Index: http://www.cdc.gov/obesity/defining.html

38 National Institutes of Health: http://ghr.nlm.nih.gov/glossary=bodymassindex

39 For some ideas, see: http://www.creditloan.com/infographics/how-the-average-consumer-spends-their-paycheck/

40 Statistical tests, if properly applied, first investigate whether the underlying assumptions of the techniques being used are met. If not, conclusions from the tests are not valid.

41 Floating point is a method of representing numbers with decimals whose place can move (float) so that the numbers can be represented in a fixed number of significant digits, called the mantissa. The locations of the decimal points are held separately, in exponents.

42 Selected Writings of Gertrude Stein, Vintage, 1990.

43 See Sheena Iyengar, *The Art of Choosing*, Twelve, 2010, and several publications of Angelika Dimoka, Director of the Center for Neural Decision Making at Temple University and her colleagues at the Center.

44 See Sheena Iyengar, op. cit.

45 Source: www.drugs.com

46 Related issues to ponder: Why are ads to the general public for drugs available only by prescription allowed at all and what effect might they be having on the costs of medicines?

47 Reported in working paper Social Science Research Network (SSRN) abstract 1814986, April 22, 2011, *Strength in Numbers: How Does Data-Driven Decisionmaking* (sic) *Affect Firm Performance?*

48 Ibid.

49 Let the reader beware.

3

Connections—the Internet, the Web, and the others

Give a person a fish and you feed them for a day; teach that person to use the Internet and they won't bother you for weeks.[1]

Getting information off the Internet is like taking a drink from a fire hydrant.[2]

The Internet and the Web, while not the only widely used information communication methodologies, certainly are in the forefront. Because of that, it's a good idea to delve into the technology behind them so we have a good understanding of how they work and some of the implications of their use. Afterwards, we'll take a look at a few other connections technologies, especially the considerably popular cellular telephony.

In the beginning—a little shiny ball

On October 4, 1957, only a little more than fifty years after the Wright brothers flew the first manned self-powered fixed wing heavier than air machine,[3] the Soviet Union launched Sputnik 1, the first artificial object ever to orbit the earth. About the size of a basketball with four wispy antennas streaming back from it, the silvery Sputnik satellite was placed into a somewhat elliptical low-earth orbit that took about 96 minutes to complete.

Astonishingly, the shiny orbiter could be seen from the earth. More than that, it spoke to us, emitting a beeping radio signal that could be heard as the tiny traveler came into view. Sputnik's batteries died after only 22 days, but it journeyed on in silence for another 64, remaining in orbit until January 4, 1958, when it plunged to earth in a fiery death.

To say that Sputnik was cause for concern, especially within the governments of the United States and its allies, is a colossal understatement. The *cold war* was in full swing at the time and the launch of Sputnik provided much fodder for terrifying visions of destruction raining down on the US and other free world countries from untouchable space. Then too there was the matter of pride—US technological hegemony was seriously threatened.

President Eisenhower,[4] not one to shy away from peril, created the Advanced Research Projects Agency (ARPA) in 1958 as an important step in meeting the challenge. ARPA,[5] an adjunct of the Department of Defense, was charged with overcoming what was perceived as an alarming technological and military lead of the Soviet Union. So began the dramatically named race for space, accompanied by an escalation of the similarly named nuclear arms race.

The cold war was a prolonged period of tense, contentious relationships principally between the United States and the Soviet Union, though most Western countries sided with the US and most Soviet allies with them. It began soon after the end of World War II (1945) and lasted until the fall of the Soviet Union (1991). Although it included economic and political dimensions, a major component of the cold war was the threat of it turning into hot nuclear war, which prospect thrust the military of both countries into prominence. The frighteningly real possibility of global nuclear annihilation shadowed everyone's thoughts.

Not simply focused on satellite threats,[6] ARPA soon realized that a critical defense weakness was the inability of the armed forces to maintain communications during conflicts. The computer systems and communications networks the military relied on within and between its branches were not fully interoperable. Further, communications were vulnerable to blackouts should key parts of the networks be compromised. That had to be addressed.

What was needed was a robust computer-based communications system that could interconnect independent mostly mainframe computers and also could deal with a complicated communications picture in which many incompatible self-contained networks were in play. Importantly, the system had to be able to keep the military's communications flowing in the face of a variety of attacks and outages. To solve these problems, ARPA created, charged, and funded the ARPANET project.

Some sources relate that the ARPANET was to be a computer communications network specifically designed to function in the event of a nuclear war, but that was not the case. After all, in a nuclear holocaust, there would be nothing left in the world to function at all. There was a small study done at the time by the RAND Corporation, an original contractor to ARPA, that sought a voice system design able to function long enough in the face of a nuclear attack to allow a counterattack.[7] It was commissioned by the Air Force and had nothing to do with the ARPANET. Perhaps that's how the rumor started. It's a good example of misinformation flowing around the Internet, as noted in Chapter 2.

A modest start—the ARPANET

The ARPANET project incorporated the ideas of a series of memos written by J.C.R. Licklider[8] while he was at MIT, describing what he called a Galactic Network, and the work of Len Kleinrock[9] and Paul Baran,[10] who developed methods for packet switching, a foundation communications methodology of the ARPANET.

Packet switching, a tremendous leap in efficiency, was very different from the so-called connection-oriented networks typified by the analog-based telephone system that were in place at the time. Besides voice phone calls, telephone networks were commonly used to transmit data because, for all practical purposes, they were the only game in town.

Telephone networks had the advantage of providing widely available consistent connection services. The problem was, they were designed to carry analog voice at a time that considerably predated digital communications. As a result, they were particularly poor at digital transmission on the important dimensions of speed, efficiency, and robustness.

A telephone connection for digital transmission is inefficient because a circuit cannot be used by anyone else while it's connected, even if no data are being sent. Since computers typically transmit data in bursts rather than continuously, there's a lot of idle time on the circuit that's not available for other transmissions. It's slow because the transmission capacity available on an individual circuit is severely limited. And since the connection, once established, is the only one that can be used for the duration of the session, if it's disrupted, transmission ends.

> **Why is it called packet switching when it's frames that are switched?**
>
> The various network processes involved in communications are separated into a hierarchy of functional groups called layers, each of which adds some control data to the message segments that eventually become frames. The segment at the layer that adds routing information is called a packet. That's followed by the layer that adds data to create the frame, which is what is transmitted. But since the networks use packet data to determine routes, it is called packet switching.

Packet switching addresses those issues directly by taking advantage of the characteristics of digital traffic and the many alternative paths from sender to receiver that digital networks can provide. Information to be sent over a packet-switched network is first parsed into small segments called frames, each of which can travel from sender to receiver independently over many possible routes. At the receiving end, the separate frames are reassembled into the original message. Route redundancy means congested paths and broken links can be bypassed, so a communication isn't disrupted. And since frames of different messages, even from different sources and with different destinations, can be intermixed on any link, what otherwise would be idle time on a link is largely eliminated during busy periods. That's the operational methodology that was the basis for ARPANET's design.

The newborn ARPANET debuted in September, 1969, when a team at the Network Measurement Center at the University of California at Los Angeles coupled an SDS Sigma 7 computer to a connection device called an Interface Message Processor, thus becoming the first node on the ARPANET network. Soon after, computers on the campuses of the University of California at Santa Barbara, Stanford University, and the University of Utah were connected.

Slow progress, then boom—the Internet

From that humble beginning, the Internet ensued. Its development over time grew out of contributions from other networks based on the ARPANET connection and packet-switching concept that emerged over the next decade. Some of the more important of these:

- CSNET (Computer Science Network) linked the computer science departments of many universities that otherwise didn't have a connection to the ARPANET. CSNET was the precursor of NSFNET.
- NSFNET was initially funded by the National Science Foundation. It served as a backbone network in the early stages of the Internet, providing the pathway that enabled many autonomous networks to be interconnected.
- USENET, a popular network promoted by AT&T, was based on the UNIX operating system, which became the foundation operating system of the Internet.

As these coalesced and other networks, protocol and control software were added, the Internet was formed.

After all this, you may be wondering when the Internet actually began. In fact, there is no date on which we can say a switch was flipped and the Internet sprang to life. Instead, the Internet was marked by a progression of precursor developments, eventually becoming widespread and foundationally unified enough to deserve its own name.

> A *backbone network* provides a common pathway to interconnect other networks. An *autonomous network* is self-contained and can operate independently. The autonomous networks of the Internet are based on what is called an IP addressing protocol that provides a consistent method and basis for uniquely labeling, finding, and routing to interconnected networks and nodes. That way, these networks can be interconnected even if the individual networks are directly incompatible.

The turning point is considered to be January 1, 1983, when NCP (Network Control Program), the original ARPANET networking protocol, was officially replaced by TCP (transmission control protocol) and IP (Internet protocol) version 4. At the time, TCP and IP were separate protocols. Now, somewhat confusingly, TCP/IP is also the name of a suite of protocols that includes the individual TCP and IP among others.[11]

IP handles node-to-node network communications where the nodes are not directly connected. It provides what is called *best effort service*, which does not guarantee packet delivery. Its primary concern is addressing and routing datagrams.

> *Nodes* are devices connected to the Internet. That means computers. It also includes switches, routers, and other network equipment, all of which are computer based.
> A *datagram* is a fully independent frame that carries complete addressing information with it.

TCP is connection oriented, somewhat analogous to telephone circuit switching. The difference is that with TCP it is the path that's fixed, not the circuit, so the path's links are available to other packets as well. TCP guarantees end-to-end packet delivery and correct ordering of the frames of segmented messages. As an overlay on IP, it can provide reliable delivery for datagrams, thus overcoming the limitations of IP's best effort operation.

Despite the January 1 turning point, ARPANET was still the extant network after the move to TCP/IP took place, not being officially retired until February 28, 1990. But from the changeover point, growth in both the size and number of interconnected networks progressed rapidly, encompassing networks in other countries as well as in the US. Somewhere along the line, most likely in the mid to late 1980s, the name "the Internet" popped up to describe this burgeoning global system, a name that has stuck ever since.[12]

So what actually is the Internet and how does it work?

Simply put, the Internet is a system of interconnected autonomous networks. By virtue of this interconnectivity, the Internet can function as if it's a single consistent network even though various of its individual networks are not compatible.

Functionally, the Internet is a transportation system for moving digital traffic from one network node to another. It doesn't care what the bits are that comprise that traffic. It is only concerned with moving them point to point in the journey from original source to final destination. As a comprehensive transportation system, the Internet carries bits that can represent numbers, characters, audio, voice, images, video, computer programs, and applications that recipients can use as they like, all in digital form.

Infrastructure-wise, the Internet consists of vast numbers of computers, switches, routers, cabled and wireless transmission systems, and software—the components of the mechanisms that comprise and connect autonomous networks and nodes along with the programs that run those mechanisms and the protocols under which they operate.

As is true of any transportation system, the Internet needs to know how to identify senders, receivers, and intermediaries to carry out its work. Since the autonomous networks of the Internet are independent, they are not required to use any particular addressing scheme to identify and find individual nodes.

Organizations typically use what best suits them and their technology. The result is there are many different addressing structures in play. Add to that the fact that we can't reasonably expect each autonomous network to know the schemes of all the others. Even if they did, the huge number of stored systems to sift through and the work required to interpret and transform different systems into the one at hand at each hop in the route from end to end would be so enormous as to drastically slow down the whole interconnection process. We seem to be faced with what looks like a recipe for chaos.

To solve the problem, all Internet member networks use a common two-level addressing scheme that is overlaid on whatever particular individual systems they are using. That scheme is embodied in the IP address, one part of which carries a unique logical network address.[13] The Internet only needs to deal with that part to route frames properly from network to network. Once a frame reaches the network where the destination node resides, routing to that node, which uses the other part of the address, is the responsibility of that network.

Finding the trees in the forest—the domain name system

Every device connected to the Internet needs an IP address. Websites are constantly being created and removed. Connection agreements with Internet service providers are signed and cancelled. New smartphone and VoIP Internet lines are activated and deactivated. Businesses form, existing businesses expand, and others fail. All of this requires adding, deleting, and otherwise managing IP addresses. Billions of active IP addresses and their associated names must be kept track of and the list keeps growing.

IPv4 addresses are numeric, for example 192.0.32.7, a notation called dotted quad. It would be tedious if not downright impossible for us to have to keep track of numeric addresses to visit Web sites and send emails. Fortunately, we don't have to. Instead, we use alphabetic names that are much easier to remember. They are automatically translated to IP addresses for us—a continuously active key function of the Domain Name System (DNS) that enables connection processes to run smoothly.

Alphabetic versions of IP addresses are called domain names. Every domain name and email address is globally unique and has a one-to-one relationship with a unique IP address. The DNS translates the name into an IP address that the Internet uses to route the transmissions. Since computers ultimately work in bits, even the dotted quad notation is a convenience for people. So there is one more translation, from the dotted quad into a string of bits. The translation process is called resolving the domain name.

The same process applies to email addresses. A computer program called a mail transfer agent sends email from one computer or mail server to another. These agents use the DNS to find out where to deliver the email.

At any given moment, millions of Web site visits and email transmissions are taking place. Managing all this is a monumental endeavor. Yet when we visit a site or send an email, translation happens almost instantaneously. How is that accomplished?

Globally, the DNS is an interconnected system of very high-speed servers running hierarchically linked distributed databases of IP addresses and their related domain names. The distributed structure makes the DNS a network in its own right, and a very efficient one at that. When translation is needed, the DNS searches its databases to find the IP address associated with the name and relays it back to the device in question.[14]

The name we type in a browser's address line is called a *uniform resource locator* (URL). It contains the global domain name and other information—the protocol being used, local network and server identifiers, and directories and files on the servers. To see what a URL's components are, let's start with a simple example:

http://www.myexample.com

The rightmost segment, .com, is called the *top level domain* (TLD). This particular TLD is ostensibly assigned to commercial enterprises. One of the original TLDs, the other five are: .edu for educational institutions, .gov for government sites, .net for network service providers, .mil used by the military, and .org for non-profit organizations and those that don't fit the other designations.

Early on, NATO petitioned for a .nato TLD and for a short time it was implemented. However, soon after this .int for international (intergovernmental) organizations was introduced. Subsequently NATO changed its domain name to nato.int. In any case, over time the distinguishing characteristics of .com, .org, and .net blurred. Now they are called *generic TLDs* (gTLDs).

To the left of the TLD is .myexample, a subdomain (also called the second level domain). The combination of the two, .myexample.com, is the *domain name*. The domain name is the only part of a URL that has to be globally unique, but that is sufficient to ensure that no two URLs are the same. Without uniqueness, orderly operation would not be possible.

Before any domain name can be used, it must be registered to guarantee uniqueness. The Internet Corporation for Assigned Names and Numbers (ICANN) was created as the central domain name registry.[15] As the Internet grew, four regional registries were added to share the workload: American Registry for Internet Numbers (ARIN.net); Asia Pacific Network Information Centre (APNIC.net); Latin American and Caribbean Internet Addresses Registry (LACNIC.net); and Reséaux IP Européens (RIPE.net). These, in turn, have created many sub-registries to handle the growing volume of work. All remain centrally coordinated by ICANN to assure domain name and address uniqueness.[16]

Moving one more to the left we find www, the name of a local server in the .myexample.com domain. Although www is a very widely used server name, probably because it often is an abbreviation for the World Wide Web, it is not a required name. What we call our servers is up to us, the registered domain name owners. Of course, other issues apply, such as copyright infringement, trademark protection, and poaching. If we register the domain name cheese.com and then use as a server name Kraft, creating Kraft.cheese.com, which is technically possible, we'd likely be enjoined by Kraft, Inc.[17]

Finally, the last step to the left brings us to http://, the protocol of the Web page that the URL refers to, in this case, hypertext transfer protocol. Among other things, http defines Web page formatting and actions taken in response to particular requests. The http protocol is the one most widely used by Web browsers.

Each http command is performed independently without reference to or even awareness of preceding commands—a so-called *stateless protocol*. That makes it difficult to create Web sites that interact with users beyond clicking on links. To overcome this limitation, software such as Java and ActiveX are used to provide interactive program code. That software also can be a vehicle for transmitting malware—yet another good/bad technology example.

Security is a different matter. By itself, http does not provide any. For sites of financial institutions and others that require secure transmissions (unreachable without appropriate passwords and protected from prying), special software and protocols are employed. So that we can see that those measures are active, an "s" is added, as in https://, indicating that communications are secure.

URLs can include more local information than our simple example—directories, subdirectories, and files on the local server. Let's extend the example:

http://www.myexample.com/sales/customers/demogs.mdb

We've added /sales, the name of a directory in the server www; /customers, the name of a subdirectory of sales; and /demogs.mdb, a database file in the subdirectory *customers*.

These examples are enough to get the gist of what URLs signify, but of course, they're not the whole story. We can name additional servers in the URL. The rightmost one would be the main server. Any to the left of it would be other local servers, perhaps departmental or for specific functions. These would be linked to the main server. We can have protocols other than http. We can have more directories, subdirectories, and files of various types. There are also country codes—two letter designations added to the right of the TLD (.ccTLD) and considered to be part of it.[18]

As the Web has grown, so has the demand for domain names. One result is that the meanings behind the original TLDs have become diluted because they are so broadly used. To remedy this situation, seven new TLDs have been formed:

● *.aero*: reserved exclusively for aviation-related organizations
● *.biz*: for businesses worldwide
● *.coop*: for cooperative businesses or those that support them
● *.info*: seemingly intended for information services, it's an unrestricted domain that any person, group, or organization can register a name under
● *.museum*: for museums and related associations and professionals
● *.name*: personalized domain names for individuals
● *.pro*: for medical, legal, accounting, and engineering professionals.

The first six of these TLDs are operational, but .pro is still under negotiation.[19]

Sorry, we're all out of addresses—IP revisited

IPv4 is the two-level (*network address/host address*) 32-bit scheme that supported rapid growth of the Internet. *Network address* refers to an autonomous network connected to the Internet—businesses need at least one of these. *Host address* refers to a computer on the particular network.

When the scheme was being designed, one question was how to split the 32 bits of an IP address into the two address levels. The decision was to establish partitions, called classes, for three sizes of organizations. That decision has a great impact on how many addresses of each type are available.

Class A is for extremely large organizations, each of which need a great many host addresses. Since there are so few of those businesses, not many network addresses were needed. The partition is an 8/24 bit split that provides just 126 network addresses but over 2 million host addresses per network.

Class B is for medium-size organizations which therefore need many fewer host addresses. Since there are many more of those businesses, a 16/16 bit split was assigned, that supplies 65,534 each of network and host addresses.

Class C is for small organizations which need relatively few host addresses. But there are a great many of them, so the split was 24/8. That accommodates over 2 million network addresses but only 254 host addresses per network.[20]

These classes account for 87.5 percent of the potentially available addresses. Two other classes were defined that used the rest: D for *multicasting* (from a source to multiple destinations); and E, reserved for experimentation.

As an organizing scheme, classful addressing made sense, but it has a significant drawback—it wastes a lot of addresses. The following is an example of why that is so.

When an organization applies for an IPv4 address, it receives a network address that carries with it a block of potential host addresses whose size depends on the address class. The organization creates its own host addresses within its block.

Suppose a company has 1,300 hosts. They would need either six class C addresses or one class B address. Six class C addresses can handle a total of 1,524 hosts (6 × 254), meaning 224 addresses go unused (1,524 available—1,300 needed). Since the unused addresses are in blocks associated with the company's six network addresses, they are unavailable to any other organizations because no others can have the same network addresses. So the unused host addresses are wasted.

Rather than dealing with six network addresses, the organization could get one class B address. That would be much worse. Since the class can handle 65,534 hosts, 64,234 addresses (65,534—1,300) are unused. Again, they would be unavailable to any other organization. Even with the relatively few wasted class C addresses of this example, multiplying by the millions of similar organizations results in enormous address loss.

When IPv4 was introduced, wasted addresses weren't much of a concern because the demand was small and IPv4, even with its classes, does provide a lot of addresses. But the demand for IP addresses has grown by rapidly increasing leaps and bounds along with the phenomenal growth of the Internet and the devices that connect to it, all of which need unique IP addresses. The upshot is that IPv4 will soon reach the end of the road, as demand for new addresses surpasses the number that IPv4 can accommodate.

To forestall IPv4 obsolescence, a version using *classless addressing* was implemented. Though more efficient, it was at best a stopgap measure and not a widely used one at that. In only a few months, there will be no more new IPv4 addresses to give out. The solution now being phased in is IPv6.

Addresses galore and more—IPv6

IPv6 has been around for about a decade, though not much used until recently. With a 128-bit address sequence instead of the 32 of IPv4, it provides a huge increase in the number of addresses available. It also allows for more than two levels in the addressing hierarchy, making improved routing possible.

Its design accomplishes several other major goals as well: faster, more efficient packet transport; better authentication to assure messages are from whom they indicate they are; additional privacy protection to make communications more secure; quality of service functions for priority transmissions; and guaranteed performance to provide better service for transmissions like audio and video streams that require high-speed steady flow.

IPv6 replaces the dotted quad with eight 16-bit segments separated by colons. Each segment is written in hexadecimal rather than decimal notation.[21] Here is an example:

A1B9:CC5F:000D:0037:FF0E:3945:0000:2A4D

As with IPv4, this notation is a human convenience; the addresses are ultimately converted to bits.

Changing over to IPv6 is complicated by the fact that IPv4 equipment can't handle IPv6 addresses. The networks and infrastructure of the Internet can't be changed overnight. Gradual cutover has been going on for some time, though lately the changeover has been speeding up in realization of the impending end of IPv4 addresses. To make cutover more flexible, three methods are available that permit functioning in mixed IPv4/IPv6 environments—dual stack, tunneling, and translation.[22]

Connecting to the Internet

Connecting to the Internet means connecting to a service provider. The Internet's service providers are arranged in a quasi-hierarchy. We connect at the bottom level, to one of the *Internet service providers* (ISPs). The rest of the hierarchy is internal to the Internet.

Regional service providers (RSPs) are one level up. At the top are *national service providers* (NSPs), many of whom are also *international service providers* (IISPs) that connect to other countries. Each level up has faster paths with greater capacity. The NSPs form the Internet backbone, the highest speed greatest capacity pathway.

Although the basic configuration is hierarchical, it actually is an interwoven structure rather than a strict tree—some ISPs can connect directly to each other, as can some RSPs and NSPs. Most businesses also connect via an ISP, though some very large ones connect directly to an RSP, in effect acting as their own ISP.

ISPs offer a variety of connection types. Dialup, the slowest and least expensive, uses the landline telephone system. Cable TV companies are the most common ISPs, offering several speed and capacity options at various costs. Once you have an account with an ISP, the connection is made automatically by software operating in the background when you start your computer. Then you are ready to use the Internet.

Consumers and providers—the network neutrality debate

At its beginning, the Internet was designed as an open access non-discriminatory transport system to move digital data of any and every type from one end-point to another without consideration of the data themselves. That describes a principle of operation called network neutrality, or just net neutrality.

In simplest terms, this means that content carried by the Internet is treated equally by all access providers regardless of content source, type, or ownership, and that access to legal content is not proscribed. That being the case, Internet performance and delivery quality would be independent of the content we seek or send, the sites we visit, and the Internet service providers we use.

While net neutrality was easy to accommodate in the early years of the Internet, it has become a debatable, contentious issue that periodically reaches public attention. The basic arguments are the following.

Network owners, operators, and service providers—independent ISPs, cable companies, and telephone companies—favor modification, if not elimination, of net neutrality constraints on their businesses. After all, the Internet of today is far different from its early days. As such, neutrality has substantial drawbacks that didn't exist then. Their main arguments:

- Internet operators and access providers need to be able to manage their networks efficiently. Some transmissions by their nature are very demanding of network capacity. Some sites generate disproportionately more traffic than others, thereby accounting for greater network loads. Since networks have finite capacities, if all transmissions are treated equally, traffic from heavy loaders will compromise service to the others. So it is logical that service be subject to control at times when network performance may be degraded, whatever the load source.
- Improving network delivery capability and capacity—building out and updating the infrastructure—is vital. It is only fair that heavier users pay more, either in dollars or by having their usage governed, so as not to put excessive demands on the systems. Pricing and tiering structures are typical allocation constructs that ensure fairness and that should apply to Internet service just as they do to other types of services. Without use-related fees or governance, the burden falls unfairly on lighter users and compromises operator ability to provide top-notch service.

Network content providers support net neutrality. For good reason, they say, the Internet was conceived as and still is a public system. Just look at the way it has grown and flourished. That wouldn't have happened without neutrality. Their main arguments:

- Neutrality prevents unfair competitive practices. Any move away from neutrality is discriminatory. Large corporations can afford to pay high fees for fast and high quality transport, but small businesses, especially startups, cannot. That puts them at a considerable competitive disadvantage. Instead of competing on a level playing field, they will be relegated to lower speed routes, have access limited or transmissions delayed, simply because they do not have the clout or financial ability to afford higher fees for better service. Since competition drives innovation, the result will be lethargic oligopolies or monopolies, to the detriment of innovation and creativity.
- Without neutrality, access providers will be able to regulate flow based on information source and content. It is increasingly common for ISPs, cable, and

phone companies to also be content providers. They have a great incentive to give preference to their own traffic and that of affiliates and to degrade or block competitor transmissions or any content they view as antithetical to their positions. Access restrictions will inevitably push users towards some sites and away from others.

Once you get to details, definitions become fuzzier and arguments nuanced. For example, ISPs charge different rates to consumers for different service levels—dialup and various broadband speeds—and that's not considered to be anti-neutrality. But that's different from an ISP deliberately favoring or disfavoring particular providers of content or services like Internet telephony (VoIP) or streaming video, which means unequal treatment.

The regulation dilemma   Whether or not net neutrality is to be maintained, the question becomes who or what sets and enforces the rules for either outcome? The Internet is global, but laws and regulations are national. While international agreements may be possible, they have to begin with the interpretations and preferences of individual nations.

As the primary regulator of communications systems in the US, the Federal Communications Commission (FCC) would seem to be the logical agency to regulate ISPs in the same way as it regulates telephone companies. The latter authority stems from its mandate to oversee *common carriers*. But about a decade ago, the Internet was classified in the US as an information system, not a communications system, hence not a common carrier and not subject to FCC regulation.

Originally, a *common carrier* was one publicly engaged in transporting passengers or goods for set fees, without discrimination. In the US in the 1800s, that primarily meant railroads. Later, truck, bus, train, ship, plane, and pipeline transport were added.

In 1910, Congress extended the definition to telephone and telegraph companies, reasoning that carrying messages was analogous to carrying people and goods.

The Communications Act 1934 codified common carrier definitions and requirements, and created the FCC as an independent overseer and rules enforcement agency.

That classification conflates the Internet with the Web and other services it carries. The Internet itself is a content carrier, not a content provider. Web sites, voice services, audio and video suppliers are the providers. Logically, the Internet has much in common with railroads and telephone systems. From that viewpoint, it is a common carrier communications system. If defined as such, it would fall under the purview of the FCC.

> In the purely legal dimension, Internet carriers seem presumptively to be common carriers. As the legal and historical surveys demonstrated, the principal legal test for whether an entity is a common carrier is whether it has held itself out to serve all indiscriminately, and most Internet carriers seem to do so. Additionally, Internet carriers seem to exhibit at least some of the public aspects which have accompanied the imposition of common carrier duties, such as . . . the manner in which the Internet has become an essential aspect of commerce and communication for many people and industries.[23]

Operating from those perspectives, in 2008 the FCC told Comcast to stop slowing down and limiting traffic from peer-to-peer sites, which it had been doing. Comcast's claim was that the huge loads presented by these sites, especially from BitTorrents,[24] made limits necessary to keep their networks operating properly.

Comcast brought suit against the FCC, saying that based on the Internet's classification as an information system, the FCC had no authority to make such a demand. On April 6, 2010, a federal appeals court, citing that same classification, ruled that regulators, the FCC included, could not ban service providers from blocking or slowing specific sites even if content from those sites was overloading or congesting the networks.

There is some but by no means universal consensus that the Internet is meant to be an open universal platform for communication and innovation. The court's ruling, based as it was on classification, affirms that stance. However, should the Internet be reclassified in the US as a communication system, then that would give the FCC regulatory authority. Still, should that come to pass, it is doubtful that any of the neutrality pro and con arguments would change. The debate rages on.

What do you think of this? *Openness is in the eye of the beholder*

In September, 2009, FCC chairman Julius Genachowski presented a plan to keep the Internet open (i.e., neutral). In a nutshell, his relevant ideas were that ISPs: can't preclude users from accessing legal material, services, or applications; can't discriminate against content providers; and have to make their network management practices public. He also allowed for reasonable network management based on conditions in the networks, as long as that did not conflict with keeping the Internet open.

In August, 2010, Google and Verizon Communications presented a joint open Internet proposal to congressional members considering net neutrality legislation. The companies agreed with the FCC's non-discrimination imperative and that there should be "no paid prioritization over the Internet." However, they also added two requirements. One, that carriers could provide differentiated online services that are "distinguishable from traditional broadband Internet access services" and could offer them exclusively. Two, that carriers could sell dedicated capacity to customers over their dedicated networks that are not part of the Internet—for example, Verizon's FiOS. (See their full proposal at http://www.scribd.com/doc/35599242/Verizon-Google-Legislative-Framework-Proposal)

On the go—bringing the Internet along

Less than a decade ago, getting a reliable Internet connection meant using a desktop or stationary laptop computer. Access on the go, whether to link to a Web site, to make Internet phone calls, or for other Internet-based applications, was problematic at best and expensive as well.

Two major impediments to the growth of mobile computing and communications were lack of widespread reliable wireless connectivity and suitable handheld devices. The substantial underserved demand for mobile communications was clear, however. That led to rapid spread of improved wireless communications infrastructures and remarkably capable mobile devices. The result—strong growth in on-the-go Internet and Web traffic. That has added two terms to the vocabulary—the *mobile Internet* and the *mobile Web*.

The terms should not be interpreted to mean that there is something different about the Internet and the Web when accessed from a mobile device. The Internet is the Internet and the Web is the Web regardless of how and where they are accessed. On the other hand, *mobile* is not a trivial distinction, because mobile operations face more difficulties than do wired networks.

Wireless connectivity, moving about or not, is not in the same reliability and continuous uninterrupted service league as fixed-location wired connections. Wireless transmissions operate over small, defined frequency ranges of the electromagnetic spectrum that have much less bandwidth (capacity) than wired connections. In addition, wireless reach (distance) is quite geographically constrained. To maintain connectivity when a user crosses a reach boundary, a handoff from one wireless transmission tower to another has to take place. If the new location's capacity is saturated, the

> The *electromagnetic spectrum* describes radiated energy in terms of frequency and wavelength. For ease of reference, particular spectrum ranges are given names. At one end is very low frequency long wavelength radiations (radio); at the other end is very high frequency short wavelength radiations (gamma rays). The upper end of the radio spectrum and the microwave range just above it are commonly used for wireless transmissions. Infrared, just above microwaves, is used for communication over fiber optic cables.
>
> (Want to learn more? Visit: http://www.space-today.org/DeepSpace/Telescopes/GreatObserva tories/Chandra/ElectromagneticSpectrum.html)

connection will be dropped. Service also will be lost when entering a "dead zone" where the carrier's wireless signals can't travel. Interference from other wireless transmissions and electromagnetic radiations present difficulties as well.

The relatively small form factors (screen size, virtual or actual keypad) of mobile devices like smartphones and tablets trade the viewability and input simplicity of laptops and desktops for light weight and carrying ease. Then too, small size devices mean small batteries that need frequent recharging. Some applications, like streaming video, are power hungry. But at least they happen when consciously invoked. More tricky are applications running in the background even when not used and so go unnoticed as they sap battery energy—for example, GPS, WiFi, and Bluetooth.

Another sort of problem comes from mobile access being troubled by inter-operability issues stemming from the lack of standards. Different mobile operating systems, carrier access and transmission protocols, browsers, animation and streaming software formats, and device platforms, many of which are not compatible, mean that for cross-platform communications to succeed, background translations must be carried out. That slows down the connection process and can interfere with downloads and streaming applications.

The mobile Internet and the mobile Web are growing rapidly, which itself can be problematic as demand approaches capacity. Yet despite mobile's problems, its increasing popularity indicates that on-the-go will surpass fixed-location access in the not too distant future.

What do you think of this? *The new 80/20*

A long-standing rule of thumb called the Pareto Principle holds that 20 percent of X accounts for 80 percent of Y—for example, 20 percent of customers make 80 percent of purchases. As a rule of thumb, it's not meant to strictly hold true, but it often has given a sense of representative proportions. When it comes to wireless bandwidth, different ratios are becoming apparent. Recent surveys have found that 90 percent of wireless bandwidth is consumed by 10 percent of the mobile users and that just 1 percent of them account for 50 percent of available bandwidth (*New York Times*, "Top 1% of Mobile Users Consume Half of the World's Bandwidth, and Gap Is Growing," January 6, 2012).

As smartphone penetration grows, wireless bandwidth demand will grow as well. Will bandwidth saturation lead to increasingly expensive data plans to parcel out a limited resource? If so, will the ratios even out or skew based on income levels?

The World Wide Web

A journey of a thousand sites begins with a single click.[25]

When I took office, January of 1993, only high energy physicists had heard of [the World Wide Web]. Now even my cat has its own Web page.[26]

We've rather casually mentioned the Web here and there—now it's time to see what it really is.

It began with a proposal   In March 1989, Tim Berners-Lee presented *Information Management: A Proposal*, to his colleagues at the European Organization for Nuclear Research (CERN). It outlined a global system for managing and transferring information over a complex internetwork via hypertext linking. The proposal was a document, not an implementation, but it is taken as the date on which the World Wide Web was born.

It may seem odd that a nuclear research agency was the place where the Web began, but not when you consider that descriptions of the great number of developments produced there and the seemingly constant flow of people in and out of CERN made it difficult to track and find particular documents. Many were ostensibly lost, whether actually missing or just not locatable. That frustrated Berners-Lee, who came up with his proposal as a solution. Working with his colleague Robert

Cailliau, the scheme was improved upon, resulting in a product called *Enquire* in 1990, in essence a server-based linked and searchable document system.

The first Web server was created by Berners-Lee the same year—he called it httpd—and the first client—called WorldWideWeb, which incorporated a hypertext browser[27] and editor. At first, Web servers were available only to a select few physicists, but in the summer of 1991 a CERN package was made available over the Internet to high energy physics researchers that included the browser, server software, and a variety of support programs for sites to create their own servers.

By the end of that year, the first US Web server was activated at the Stanford Linear Accelerator Center. From that point, growth was rapid.

> By the end of 1994, the Web had 10,000 servers, of which 2,000 were commercial, and 10 million users. Traffic was equivalent to shipping the entire collected works of Shakespeare every second.[28]

So what actually is the Web?   The Web is a vast collection of servers organized into Web sites—collections of files called pages that are stored on the servers. A Web site page can have links to places within itself, to other pages on the same Web site, and to other Web sites. The link structure is called hypertext. The links, also called hyperlinks, are actually addresses that take us from page to page, site to site. That makes traversing the Web a straightforward background process.

The Web runs on the Internet but it is not synonymous with the Internet. We use the Internet to go to Web sites much as we use any transportation system to get to our destinations. In the case of the Web, the Internet is the transportation system and the Web sites are the destinations.

The Web is based on a *client–server* model to handle interactions among software processes. In a nutshell, clients request services and servers provide them. When we want to go to a Web site, our browser software (*client*) requests a Web page from the site's server software (*server*).

The emphasis on software is intentional. Hardware devices can act as clients or servers—they do not need to be dedicated to task. It actually is the software that provides request–response functionality. It is common for Web servers to be dedicated machines, but even then, if one server requests files from another, it is acting as a client.

The Web 1-2-3   The Web we are most familiar with is the enormous array of independent sites serving pages on request. That being the first implementation, it was retroactively called Web 1.0. Over the last six or seven years, applications have changed significantly enough to justify a name upgrade, to Web 2.0.

The latter began with affiliations that bring into play other sites or content, providing combined services sometimes called mashups. For example, check out a hotel and a MapQuest or Google map pops up with directions, lists of local restaurants, entertainment, and other related information, not to mention advertisements.

Next came greater user-focus for easier sharing of all sorts of information, collaboration, participation, and creation of virtual communities. That provided the mechanisms for wikis, blogs, social networking, podcasts, content aggregators, and hosted services like online data storage, software run remotely, and provision and maintenance of your own Web site.

Web 2.0, then, is not a change in the technology of the Web or the Internet, but a change in the way Web applications are constructed—the way developers design Web sites—to make them much more interactive than Web 1.0's viewing of static pages. Even so, Web 2.0 has a noteworthy shortcoming—it doesn't understand us.

When we search for information, the results are based on keyword combinations and phrases, but context and interpretation are absent. The intention or meaning of what we're looking for is not considered. The next step, then, is to add that capability, producing Web 3.0—the *semantic Web*, a term coined by Berners-Lee. That hasn't happened yet, but when it does, our requests will be interpreted to glean the essence of what we seek. It will understand us.

Now, if we want to buy an LCD TV, we could search for those words, for brands, for sizes, and the like. But with a semantic Web we could say *Let's see a comparison of LCD TVs including reviewer ratings, reliability, power consumption, and availability ranked in order of those terms.* If we are thinking about taking a vacation, we could ask for *information on rentals at seaside resorts with accommodation for two adults, two children, and pets, within 50 miles of an airport in France, Spain, or Mexico, with costs and comments, availability in June, and oh yes, include color photos.*

As it happens, the semantic Web is only one idea, although the most popular one, of what 3.0 will be. Some take it to mean virtual worlds, merging of the real and the virtual, and even the Web creating information on its own. Others confuse the Web with the Internet and other connection mechanisms by seeing 3.0 as having features like wireless high-speed access from anywhere and Web connections from all sorts of devices that can then communicate with each other. Still others consider some of 3.0's more esoteric features, like virtual 3-D communities, as being Web 4.0.

Behind the portal   Most Web sites have a particular focus—an automobile company, a designer clothing brand, a sports team, a charity—just about any kind of business or service there is, and so are end-point destinations. The main page of a Web site is called the homepage. These days, it's common for a focused homepage to act as a *portal*—an organized page of information and links that go deeper into the site to its various features and content. There may be a few links to other sites, but those almost always are affiliated with the originating site in some way.

There is another kind of Web site that is not focused on a particular business or service organization, but instead is an organized entryway into a large variety of services and other Web sites. Instead of being an end-point destination, it is a starting point. It's called a *Web portal*.

Web portals are quite different from other Web sites. Web portals allow you to customize what you see and where you see it on the page, and to add or delete features. Other Web sites do not. Web portals have a great many links and they go primarily to independent sites that do not have any particular connection to each other or to the portal. Typically, much of the information on a Web portal changes quite frequently, while a lot of the information on non-Web portal Web sites is more stable.

To get a better understanding of the difference, let's look at what's on one popular Web portal. What you see on the page includes:

- Several pictures related to current interest stories, with links to sites with additional information
- Current values of the Dow, Nasdaq, and S&P and a search box to get a quote on a specific stock
- Links to several editors' picks for current news stories
- Links to sites with information about all sorts of items, organized in a tabbed list—Must See; News; Getaways; Life/Style; Celebs & Gossip; Sports; Love; LOL; Local; Movies; Travel
- Search box for shopping
- Box for general searches
- Links to Facebook, Twitter, email
- Links to weather reports, forecasts, and conditions in any part of the world
- Several ads with links to their Web sites.

And that's just the beginning of what's on this portal (www.MSN.com). Take a look.

Business- and service-focused Web sites follow the Web portal model for their own sites to make it easier to find the features and services offered. They begin with a homepage that shows all the things you can find by linking deeper into the site itself. They could be called business portals or corporate portals, but they are so common now that they are usually just called Web sites.

Here's some of what you'll find on the homepage of one of them:

- Links to information about items by category—women, men, juniors, kids, jewelry, handbags, shoes, for the home, kitchen, dining, furniture, bed & bath
- Notice of a sale
- Notice of specials
- Link to the store's catalog
- Store finder, locations, and hours
- Link to customer services
- Link to the store's credit card.

And again, a lot more. But this portal, www.macys.com, is all about the department store and the pages linked to, with a couple of exceptions, are within the site.

Another type of Web site also follows the portal model to provide links to a wide variety of information, but since all the information is located at the site, they are more akin to regular Web sites than Web portals. Examples of those are Wikipedia and YouTube.

Because the portal model is becoming ubiquitous, some call any homepage a portal. More useful is a trend to label them by focus. So we find classifications like government portals, news portals, industry portals, stock portals, and many others. Some of these require membership, account creation, and login/password to gain access, something that no Web portal does.

At this point, a clarification is in order. As different Web sites copy the features of others and create their own as well, there is some overlap between what we may call pure Web portals and other Web sites. You will find some links on Web portals that go deeper into the portal site itself and some on "regular" Web sites that connect to independent sites outside of their own realm.

So far in the latter case, though, there is some relationship to the originating site— an automobile Web site might have a link to the site of an independent tire manufacturer whose tires the auto company uses and whose service departments sell; a store may have a link to the site of an independent perfume company whose perfumes they carry.

A Web site for any business—or for you

If you have a Web site, it makes your small business look big.[29]

Can you think of a large business that doesn't have a Web site? That would be a rare find. In fact, as we more and more routinely turn to the Web to find a business or service, Web sites for just about any company are becoming common, especially since now, size doesn't matter. A Web site is within the reach of any business or person who wants one, at least theoretically.

To have a Web site, two things are needed—a server to hold and provide access to the site's pages, and a connection to the Internet. Large firms are likely to have their own servers and connections. Otherwise, outsourcing to a *Web-hosting service* is the answer.

Hosting services handle Web sites from the very simple to the quite complex, with associated fees. They can provide Internet connectivity, Web site design, creation, and maintenance, and dedicated site space on their own servers or on space they contract for on other companies' servers. Most can incorporate hosted-site databases linked to the site's pages.

Websites for corporations span the gamut from single pages for very small businesses that just want a presence on the Web to raise their profile, to multiple-page sites for companies that are looking for significant customer–business interaction— and that includes business-to-business as well as business-to-consumer. They can incorporate interactive forms and features, searches within the site, site maps, animations, audios, videos, and deep and external links—in other words, any of the features you'll find on any Web site.

> **If you build it they may not come**
>
> Simply having a Web site does not mean instant business or recognition. The site could easily go unnoticed. So first, you have to create awareness and curiosity. That means coming up with creative ways to drive viewers and potential customers to the site. Once they are there, they need good reasons to utilize the site, make purchases, and come back again and again.

It's not just businesses that can have a Web site—so can anyone who is able to contract for one with a Web-hosting service. Personal Web sites are likely to be simple one-page fairly static affairs but they also can be more elaborate. Many ISPs offer free hosting for simple sites as part of their connection packages.

And the others

Using the Web requires using the Internet, but using the Internet does not require using the Web. The more popular applications of the latter are:

- *ftp* (file transfer protocol), to upload and download files to and from ftp servers by transporting them over the Internet (or in fact, other-than-Internet networks).
- *email*, which can be invoked using a Web browser but doesn't have to be. Some corporate email systems do not use browsers and most ISPs offer an email service that is not browser dependent.
- *VoIP* (voice over Internet protocol), which converts conversations to data frames that travel over the Internet. Cable companies offer VoIP as part of their Internet connection service. Skype and Google Talk are examples of independent services.
- *IPTV* (Internet protocol television), which works the same way for TV as does VoIP for voice. AT&T's U-verse is one of many examples.
- *IM* (instant messaging) and many other chat variations for real-time texting. AOL, Microsoft, and Yahoo have IM products. Unfortunately, they are not compatible.

There are important IT communications systems that don't use the Internet at all. Satellite phones provide voice service in areas where it otherwise doesn't exist, but they are very expensive and so are not a resource for most people. A prime example of a relatively inexpensive alternative is *cellular telephony*.

Apart from their wild popularity in most of the world's countries, cell phones have played an important role in personal contact and information dissemination where there are few if any options. We've also witnessed the successful use of cell phone communications to coordinate and report on some recent popular uprisings against authoritarian regimes. (Chapter 4 "That is to say—free expression and privacy" has some examples.)

Because of the ubiquity and importance of cellular telephony technology, a closer look is in order. A cell phone is just a low-powered data transceiver (transmitter/receiver), but one that's supported by a complex integrated infrastructure. To use a cell phone, you have to be in contact with a stationary *base station*. Each base station has transmission and receiving antennas plus communications equipment linked to its nearest neighbor stations. A base station covers a particular geographic area called a cell, hence the term cellular. Base stations are connected to, controlled, and coordinated by *mobile telephone switching stations* (MTSOs). They set up and terminate calls, provide links to wired landlines, to the Internet where contracted for, and keep usage and billing records.

What do you think of this? *Cell phones do the job*

Sub-Saharan Africa has many areas where the Internet is essentially unavailable and where landline phones are rare. But they also have the most rapidly growing cell phone market in the world. Millions of Africans use cell phones for applications as diverse as texting, transferring money, checking weather and produce market forecasts, and selling and buying goods and services. Here is one example:

Neither landlines nor Internet access are common in Kenya, especially in farming areas, but over 75 percent of Kenyans have cell phones. They shop and send and receive money through their cell phone accounts. Cell phone service providers facilitate the process by serving as banks—they now hold about 40 percent of Kenyan's savings. Since 2009, Kilimo Salama has been selling previously non-existent crop insurance to small farmers in Kenya. The program works through cell phones, which farmers use to send text messages to sign up for insurance and to report losses, and which Kilimo Salama uses to send payments to farmers' cell phone accounts.

(Interested in learning more? Visit: http://opinionator.blogs.nytimes.com/2011/05/09/doing-more-than-praying-for-rain/)

Landline phones or cell phones?

In areas where landline systems are well established, it costs less to provide telephone service by wire than by cellular. Landlines are more reliable as well, though of course, not mobile. Interestingly, the picture changes where no significant landline infrastructure exists. Building out the wired infrastructure is very expensive and there may be issues as to where wires can be run. Cellular requires several base stations and minimal wire to connect them and MTSOs—a much less expensive proposition. That's why in most developing countries, cell phone service far exceeds landline coverage.

When you make a cell phone call, the system establishes a connection between your phone and the base station of the cell you are in at the time. Its MTSO checks your account and phone to authenticate you as a legitimate user. Then it tries to locate the party you are calling, whether the call is to another cell phone or a landline. Calls to a landline phone are routed through the regular telephone system. If the call is to a cell phone in another cell, the associated MTSOs have to communicate with each other. Once the call goes through, if you or the called party move to another cell, the MTSOs hand off the call to the base station in the new cell.

That's the simple explanation. Complicating the picture is the fact that, at least in the US, each cell phone carrier (Verizon, AT&T, Sprint, and others) has their own base stations, MTSOs, and communications equipment, none of which follows a single standard. They don't support the same phones or service and they use different incompatible transmission protocols, the two most common being PCS/CDMA (Personal Communications System/Code Division Multiple Access) and GSM (Global System for Mobile communication). So if you call someone who uses a different carrier, the MTSOs have to coordinate with each other and perform continuous protocol translations. That in itself can account for connection delays, interrupted service, and dropped calls.

Call quality has a lot to do with cell size. That depends on several factors, the most prominent of which are population density, terrain characteristics, and building heights and concentration. Cell phone systems use microwaves to transmit calls. Microwaves travel in straight lines, meaning they require line-of-sight to the towers. So the more tall buildings or high ground in an area the smaller the cell has to be to maintain line-of-sight. As for population density, each component of the systems has a finite capacity, so when there are a lot of users in an area, smaller cells are needed to distribute the load.

> *Line-of-sight* means that from where you are, you can "see" the receiving antenna. Fortunately for communications, microwaves can pierce objects that are opaque to our eyes and under certain circumstance bend around or split off of objects in the way of line-of-sight, thus extending "sight."

Smaller cells mean more cells and more cells mean more base stations, hence more antenna towers. Every carrier that supplies service to an area needs its own complement of stations and towers. Further, antenna location within a cell is a critical determinant of how well the service will function. So it is no surprise that proper antenna placement is a key determinant of high quality service.

Quite often, where antennas are placed is a contentious community issue that looms large in the conflict between better service and neighborhood desires. Many community residents, wanting to preserve the characteristics of their environs, don't want antenna towers and base stations in particular locations, especially since there are multiple carriers involved. Their opponents prefer service over aesthetics, since the latter can mean poor service or no service. Carriers attempt to find inconspicuous locations or disguise antennas in ways that serve the purpose without raising objections, but that isn't always possible when technological requirements have to be met.

And so—wielding power

Measured by influence impelled by reach, speed, applications, and number of users, the Internet is the greatest disruptive information technology in the history of the world.[30] It has been a player in incredible changes, from subtle to upheavals, in almost every sphere of human endeavor and behavior. One of its more impactful outcomes has to do with wielding power.

In that quest, the Internet comes into play in two ways—how we can use it and how the few large corporations that own and run the Internet infrastructure can exert their influence. The second of these may give you pause, since no one owns the Internet, right? Well yes, conceptually, but the physical infrastructure

> The major *owners and operators of the Internet infrastructure* currently include: AT&T, Cogent Systems (multinational), Qwest Communications International, and Verizon in the US; Telstra and Optus in Australia; and Vodafone in Europe. All are public corporations. CERNET, CHINAGBN, ChinaNet, and CSTNET are major operators in China; all are state controlled.

is owned, mostly by public corporations in open societies and by the governments of closed societies.

These corporations, as RSPs and NSPs, are responsible for the operation of the upper-level Internet hierarchy pathways. Many operate ISPs as well.

In open societies, the power they wield is not directed to controlling the Internet, though we have seen arguments justifying flow-based service management in the net neutrality debate. Instead, they use their positions in negotiations with businesses that have a mission critical dependence on the Internet, especially local ISPs.

In closed societies, controlling Internet access is vital to maintaining power. Leadership's first order of business is to manage information flow by limiting Internet connectivity, blocking sites deemed inimical to the state and ensuring that only sanctioned messages are seen.

China, as a developed country tightly reigned by the Communist Party, is the prime example of this practice. The ruling party typically justifies restrictions as the state protecting the populace from insidious and diabolical forces that are a danger to the nation and its inhabitants. That has resulted in a cat and mouse game whereby the citizenry finds ways around the blockages and the state erects new barriers as they do. (This is discussed further in Chapter 4.)

Dictator-controlled under- or undeveloped societies present a similar picture with different motivations. Those autocrats also seek to wield power, but typically to usurp much of the countries' wealth for themselves, their families, and particular associates, leaving the vast majority of the populations impoverished. Yet they also benefit from the same sort of information manipulation—allowing access only to what they want known—because holding power, even by threat of or actual force, is easier when the citizenry is kept in the dark. As IT in the hands of the populace reaches beyond state control, pressure for more representation and sharing of wealth grows. We've seen some of the results in the Middle East uprisings of late. (This also is explored in Chapter 4.)

A different matter in open societies is amassing power for ourselves. We can do that by building a significant audience—a proven way to develop the extended influence needed to gain authority. The force of our positions must attract not only the like-minded but also convince others to join the fold.

What do you think of this? *The off switch*

Top executives have always had secretaries to screen calls and limit contacts. Not being reachable was taken as a sign of power, and one not available to the general public. Of course, in the pre-Internet pre-mobile phone days, we were unreachable for a lot of the time, simply by not being near a phone. But what to do when the phone was at hand? The choice was to answer it or not, though the ringing bell exerted a strong siren call to see who was phoning—the caller, or the phone system, exerting power.

We got a little more power over that urge and call screening as well when the answering machine came out. Now, we have caller ID to see who is trying to reach us and decide whether or not to answer. But we still can feel compelled to answer a call from particular parties even if we don't really want to.

Here's a way to wield the power of being unreachable—turn the phone off.

To get our message out, both we and the people we need to reach must look to the most extensive means of communication in use, the Internet, the Web, and cellular telephony. Aside from reach, they go a long way to equalizing potential use by anyone because of how little cost is involved.

To use these ITs successfully, our intended audience must have access. In any society, this has two dimensions: a connection to the service and the ability to use it. The first is a technology resource issue—owning or having the right to use appropriate equipment and connections. The second is a question of technical knowhow—how to use the equipment. This can be the key factor in the equation on the population side—the people we want to get our message.

In some areas of the globe, resource provision is actually easier to resolve than lack of knowhow, even where infrastructure funding is inadequate. It may seem that logging in and surfing the Web through computers or smartphones is a simple matter, but for many people it is not. A basic level of education and experience with technology is assumed but may be lacking, even in some parts of technologically advanced countries.

The persistent digital divide   The digital divide, the disparity between those who have available, affordable, digital communications that they know how to use and those who do not, speaks to these issues. Not being able to communicate electronically is a considerable disadvantage in everyday life, whether or not wielding power is the quest.

We've seen examples of how cell phones can close the gap to some degree, but that is only a small part of a big picture. Though the dichotomy between widespread access in open societies and little access in closed societies is quite large, living in an open society doesn't come with guarantees. It's easy to assume that access is universal, especially in highly developed nations, but in fact that is not the case.

The divide has been a contentious issue for a long time. Despite efforts to eliminate it, the divide remains stubbornly significant. A good part of that persistence owes to the rapid advancement of technology, which means that our access and services quickly lose ground if we can't keep up. That creates a different kind of divide— alongside the gap between the haves and the have-nots is a gap between the well-served and the poorly served.

Advanced societies have the responsibility to narrow all aspects of the divide in their own countries and in the poorly served ones as well. A fully communicating world will be a better place for all.

Notes

1 Anonymous.

2 Mitchell Kapor (1950–), personal computer pioneer and IT specialist, founder of Lotus Development Corp., designer of Lotus 1-2-3, one of the first "killer apps" that drove PC development, usage, and market.

3 Wilbur (1867–1912) and Orville (1871–1948) Wright, December 17, 1903, Kitty Hawk, NC.

4 Dwight David Eisenhower, former US Army five-star general, served two terms as the 34th president of the United States, from 1953 to 1961.

5 The Agency changed its name periodically, shifting between ARPA and DARPA (Defense ARPA). The last change was to DARPA in 1996.

6 The National Aeronautics and Space Agency (NASA), founded in 1958, directly focused on space exploration and utilization. Superseding the National Advisory Committee for Aeronautics (NACA), it was another governmental response to Sputnik, but unlike ARPA, it was not a Department of Defense agency.

7 The Rand Corporation, 1962.

8 Joseph Carl Robnett Licklider (1915–1990), American computer scientist.

9 Leonard Kleinrock (1934–), American engineer and computer scientist.

10 Paul Baran (1926–2011), Polish American engineer.

11 The original TCP/IP was developed by Vincent Cerf and Bob Kahn, with much input from Jon Postel. IP addressing went through three versions from the early 1960s until 1981, when IPv4 became the standard. It is still used today, but is being replaced by the newest version, IPv6. Want to discover more Internet history? Visit: http://www.isoc.org/internet/history/brief.shtml

12 There is no consensus as to who named the Internet. Vint Cerf and Bill Gates are popular choices among others.

13 Every node also has a unique physical address, but that's not sufficient for routing because it doesn't contain any location information; it simply labels the node. IP addresses are mapped onto physical addresses; they identify the individual network of the node and the node itself.

14 As you might suspect, the DNS and its operations are considerably more complex than our brief overview. If you would like more information, http://www.internic.net/faqs/authoritative-dns.html provides detailed explanations without getting overly technical.

15 The dotted quad example 192.0.32.7 is the IPv4 address for ICANN's Web site, whose URL is: www.icann.org.

16 Want to know more about ICANN and the registries? Visit: http://www.icann.org/

17 To settle disagreements, ICANN established the *Uniform Domain-Name Dispute-Resolution Policy,* which all registrars follow. For the complete policy, visit: http://www.icann.org/udrp/udrp.htm

18 There are over 240 ccTLDs. For a full list visit: http://www.iana.org/cctld/cctld-whois.htm

19 For more information, visit: http://www.internic.net/faqs/new-tlds.html. It is interesting to note that the new TLDs are not restricted to the "dot and three letters" format of the original TLDs.

20 If you calculate the number of addresses that should be yielded by a particular bit count, you'll find that they appear to be too low. That's because some bits are needed to identify a class, so are not available for addresses and some addresses are not used (for example, all 0s and all 1s).

21 Hexadecimal is a base 16 number system. The 10 values from 0 to 9 appear as is. The remaining values, from 10 through 15, are represented by the letters A through F.

22 If you would like to know more about IPv4 classful and classless addressing, address masks, subnetting and supernetting, and IPv6, see M. Barry Dumas and Morris Schwartz, *Principles of Computer Networks and Communications*, Pearson Prentice Hall, 2009, pp. 281–291. Many countries are involved in IPv6 development and deployment. For more about that, see: http://www.ipv6forum.org/.

23 James B. Speta, "A Common Carrier Approach to Internet Interconnection," *Federal Communications Law Journal* 54, no. 2 (2002).

24 BitTorrent is a distributed system for sharing files by downloading file segments from among a great number of participants. Anyone who has received a file is a source for a segment for another user. Though it works for any file, it is typically used for extremely large files like complete movies.

25 Anonymous.

26 President Bill Clinton, in a speech on the Internet for schools, Knoxville Auditorium Coliseum, Knoxville, Tennessee, October 1996.

27 A browser is a graphical software interface between us and the Web. It generates commands that are executed as we point and click, so that we don't have to enter them, greatly simplifying Web navigation.

28 http://public.web.cern.ch/Public/en/About/WebStory-en.html

29 Natalie Sequera, Claris Corporation, as quoted in: http://bohtech.com/web/

30 A disruptive technology is one that causes a dramatic shift in the way things are done and affects a great many people.

4

That is to say—free expression and privacy

A popular government without popular information or the means of acquiring it, is but a prologue to a farce, or a tragedy, or perhaps both.[1]

... now that everyone is free to print whatever they wish, they often disregard that which is best and instead write, merely for the sake of entertainment what would best be forgotten, or, better still be erased from all books.[2]

The free expression mandate

The weather in Philadelphia from May to September of 1787 was quite hot and so were the arguments among the Federal Convention delegates who were drafting what was to become the Constitution of the United States. The resulting document was submitted to the Confederation Congress for debate.[3] The Congress, in turn, sent it to the legislatures of the 13 states, which formed conventions of their own to evaluate it.

During the many debates, several legislators and other interested parties believed the draft Constitution lacked certain provisions they wanted to add, some even making them a condition of their ratification votes. Many of those ideas were rejected, but others were considered to be important. James Madison[4] put together ten amendments that encapsulated the latter. They came to be known as the Bill of Rights because they focused on the rights of the citizenry.

In 1789, Madison introduced the amendments to the Congress and they were passed on to the states as addenda to the draft Constitution. By late May 1790, all the state conventions had voted to accept the Constitution and the ten amendments, and so they were ratified. They officially replaced the Articles of Confederation, the governing document in place after the 13 colonies became independent. But the wrangling didn't end there. Questions of interpretation ensued and they continue to this day.

The first amendment has six clauses that set the foundation for the cornerstone of a participatory democracy. It reads:

> Congress shall make no law respecting an establishment of religion, or prohibiting the free exercise thereof; or abridging the freedom of speech, or of the press; or the right of the people peaceably to assemble, and to petition the Government for a redress of grievances.

Freedom of speech was, quite literally, the ability to speak freely without fear of government reprisal. Freedom of the press focused on organizations that published news, thoughts, opinions, and the like via the printed page, functions vital to keeping the populace informed. An important adjunct of free speech is noted in the last two clauses—the right to peaceably assemble and to petition the government, again without fear of reprisal.

Why the First Amendment?

The underlying principle of the First Amendment is that a free, open, democratically structured society requires, in fact demands, a well-informed public who can express their will and opinions without trepidation. That means information must flow freely in all directions, to and from all people, up and down all government levels, and by all means, without fear of coercion or retribution. So freedom of speech, press, assembly, petition, and freedom from religious imposition are explicitly stated.

Over the years the definition of speech has been considerably broadened to include any form of expression—paintings, displays, videos, audios, plays, performances, gestures, and even flag burning—that is, *freedom of expression*. Peaceable assembly has expanded from physical meetings to include virtual meetings, an important extension in the digital age.

As written, the First Amendment explicitly applied only to Congress—"Congress shall make no law"—but it has since been interpreted to cover all of the federal government. States were another matter, though, and weren't directly addressed until much later. The Fourteenth Amendment, passed in 1868, dealt with this issue among others. The first part of section 1 of the Fourteenth Amendment defined citizenship; the second part, which came to be known as the *due process clause*, has been used to assure that states did not contravene the protections of the First Amendment or the rest of the Bill of Rights.

> Section 1 of the Fourteenth Amendment: All persons born or naturalized in the United States, and subject to the jurisdiction thereof, are citizens of the United States and of the state wherein they reside. No state shall make or enforce any law which shall abridge the privileges or immunities of citizens of the United States; nor shall any state deprive any person of life, liberty, or property, without due process of law; nor deny to any person within its jurisdiction the equal protection of the laws.

Limits to protection

According to the concept of *natural rights*, certain rights are taken to be absolute, meaning that they exist for all individuals merely by fact of their being alive. Those expressed in the US Declaration of Independence, labeled as unalienable, include life, liberty, and the pursuit of happiness.[5]

The fact is, such rights are more akin to beliefs than actual rights—the Declaration states that unalienable rights are endowed to us by our creator. In practice, our rights depend on the societies in which we participate and are not absolute.

Free expression, which by logical argument may be derived from unalienable rights and, by extension, itself absolute, also is not unbounded. The challenge is to determine which expression should be protected and which should not. Drawing a clear line is easier said than done.

Many limits are contextual   Hate speech—that which attacks a person or group based on ethnicity, religion, sexual orientation, gender, and the like, induces others to make such an attack, or incites to violence—would seem to be a clear exception to free speech. Sometimes it is, but most of the time it is not. The principle of free speech is so strong in the US that the line between unprotected and protected hate speech leaves little unprotected. That someone finds particular speech odious or abhorrent is not enough to shift the line.

> The U.S. Supreme Court has consistently held that speech may not be prohibited simply because some may find it offensive. Virtually every time someone is arrested for this, assuming there's no other criminal behavior . . . the case is either dismissed before trial or the person is convicted at trial and wins on appeal.[6]

That is a consequence of the strongly held notion that the right to speak one's mind, even when the speaker abuses the government that defends that right, must be protected in a free and open society. As Voltaire put it: "I disapprove of what you say, but I will defend to the death your right to say it."[7]

Stepping over the line is a category allied to hate speech, namely hate crime—unlawful actions like property destruction, assault, murder—whose perpetrators choose victims based on the same biases expressed in hate speech or that are accompanied by manifestations of hate speech. That explains why a parade of neo Nazis waving banners and signs that are clear expressions of biased hatred is protected as lawful speech but carving a swastika into someone's door or burning a cross on the lawn of a private home is a hate crime, not protected speech.

What do you think of this? *A line too far or not?*

The Westboro Baptist Church of Topeka, Kansas, is long known for its homophobic, anti-Semitic, anti-Catholic, anti-Islamic bigoted views among others. For at least the past twenty years, they have demonstrated at various venues carrying such signs as "Thank God for IEDs [improvised explosive devices]", "Thank God for Dead Soldiers," "Fags Doom Nations," "America is Doomed," "God hates Jews," and the like.

When church members demonstrated at the funeral of Lance Corporal Matthew A. Snyder, a Marine killed in Iraq, Albert Snyder, the Marine's father, sued for "intentional infliction of emotional distress, intrusion upon seclusion, and civil conspiracy." Though he won a considerable judgment, it was overturned on appeal. Eventually finding its way to the Supreme Court, the church's right to demonstrate was upheld on free speech grounds because the pickets were on public ground, were not physically interfering with the funeral, and demonstrated peacefully.

Chief Justice John Roberts, writing for the majority, said:

> [Free speech] can stir people to action, move them to tears of both joy and sorrow, and— as it did here— inflict great pain. [But . . .] we cannot react to that pain by punishing the speaker.

Justice Samuel Alito, writing in minority dissent, said:

> Our profound national commitment to free and open debate is not a license for the vicious verbal assault that occurred in this case.

Source: http://www.law.cornell.edu/supct/html/09-751.ZS.html

Audience and purpose can make a difference   Obscenity is treated as a crime in some situations and not in others. Pornography is a free expression exception when it comes to children. On the other hand, the argument that children are involved is rarely seen as sufficient to repress free speech rights of others. A case in point is the 7–2 Supreme Court decision rendered in June, 2011. It upheld a lower court's decision overturning a 2009 California law that prohibited selling violent video games to minors, defined as anyone under 18 years of age.[8]

The California law, which the state claimed was instituted to prevent children from being desensitized to violence by playing such games, imposed a $1,000 fine on retailers who sold them to children. The majority decision of the Court in striking down the law as unconstitutional specifically cited free speech rights, saying that video games, like other forms of expression children are exposed to, were entitled to free expression protection.

Writing for the majority, Justice Antonin Scalia said

> I am concerned with the First Amendment, which says Congress shall make no law abridging the freedom of speech. And it was always understood that the freedom of speech did not include obscenity. It has never been understood that the freedom of speech did not include portrayals of violence.[9]

Another example concerns public displays of nudity that, illegal in most jurisdictions, have nevertheless been allowed as free expression when they are part of events, installations, and happenings labeled as art. One prominent practitioner of the latter is Spencer Tunick, whose installations feature large numbers of nude adults that he photographs and videos as they pose en masse in public spaces.[10]

The FCC's budget (about $339 million in 2011), is funded totally by regulatory fees, giving it a degree of independence it would not have if it depended on congressional funding. However, it is subject to congressional oversight, and Congress can overrule FCC actions and directives.

So can the type of IT and the role of the FCC   Another assortment of exceptions and regulations depends on the communications technologies entailed. The role of the Federal Communications Commission (FCC) in placing boundaries is illustrative.

The *Communications Act of 1934* created FCC as an independent federal agency to regulate interstate communication by radio broadcasting, wired telegraph, and telephone service. Foreign communication that originated or terminated in the US also fell under FCC jurisdiction.[11] Over the years, the FCC's authority was extended to broadcast television and satellite service, along with specific limited control over cable service. The particular IT in question is what determines whether the FCC is a stronger or lesser regulatory player.

Broadcast media Broadcast radio and TV transmit over portions of the lower frequency range of the electromagnetic spectrum.[12] To use a slice of the spectrum, a broadcaster must obtain a license from the FCC that allocates a specific frequency range and specifies broadcast signal power limits. Licenses also come with certain provisos concerning program content and must be renewed at regular intervals.

Program content is where it can get confusing. For one thing, license holders don't have the free speech protection of other communications media. The FCC prohibits obscene material, but makes some exception for the lesser matters of profanity and indecency.

Within limits, profane and indecent content are protected—allowed between the hours of 10 p.m. and 6 a.m. That's the so-called safe harbor time, a rather arbitrary period based on the assumption that children will not be in the listening or viewing audience during those hours.

We are again in rather blurry definitional areas. For obscenity, the FCC uses the Supreme Court's definition, which actually is aimed at pornography. One step down from obscenity is indecency, defined by the FCC as: "sexual or excretory material that does not rise to the level of obscenity." Profanity is another step down: "words that are so highly offensive that their mere utterance in the context presented may, in legal terms, amount to a nuisance."[13] All these definitions are highly subjective.

Adding to regulatory complexity, the FCC cannot censor broadcasts by vetting content in advance. Yet after the fact they can levy fines, revoke licenses or deny license renewal for violations of the content rules. (Criminal violators are subject to imprisonment.) So although

> **Why limit broadcast power?**
>
> The frequency range of the electromagnetic spectrum usable for broadcast is finite and relatively small, enough to divide over just a few dozen stations. To enable thousands of stations across the country to operate, the same slices of the spectrum have to be used by many different stations. But the higher the signal power the farther it reaches, so unless power is sufficiently limited, stations using the same slice would have to be farther apart to avoid interference. That would mean many fewer stations could be accommodated.

> **The Supreme Court on obscenity**
>
> "To be obscene, material must meet a three-prong test: (1) an average person, applying contemporary community standards, must find that the material, as a whole, appeals to . . . prurient interest . . .; (2) the material must depict or describe, in a patently offensive way, sexual conduct specifically defined by applicable law; and (3) the material, taken as a whole, must lack serious literary, artistic, political, or scientific value."
>
> (http://www.fcc.gov/guides/obscenity-indecency-profanity-faq)

the FCC cannot directly censor programs, they can do so indirectly by the threat of potential punishment. One result of this is that, in many cases, what appear to be live broadcasts are actually transmitted after several seconds of time delay so that broadcasters can zap potential violations before they are aired.[14]

Wireline telephone companies, as common carriers,[15] are another story. They have to offer service without considering caller identity or call content, and must provide clear and complete information regarding rates and billing. The FCC can regulate telephone company operations, but importantly, because they have no jurisdiction over wireline call content, there free speech reigns.

Enhancing competition in the telephone industry has become a strong FCC mandate. The idea is that competition will keep prices down, making uncensored service available to more people. They were aided in that quest by a 1984 Supreme Court decision that broke up AT&T's telephone monopoly, restricting it to long distance service and dividing local service over a number of regional operating companies, each of which covered unique sections of the country.

The breakup was only a partial fix, because the regionals, each covering different geographical areas, did not compete with each other. Also, for all practical purposes, competition from newcomers was largely impossible. To operate a telephone company, they would have to build an infrastructure—a prohibitively expensive proposition that effectively eliminated the possibility.

That was one factor addressed by the *Telecommunications Act of 1996*, a considerable enhancement of the 1934 act.[16] Among its several provisions were those intended to increase competition in the telephone industry. Infrastructure owners were required to sell capacity to newcomers on a wholesale basis, thus removing the costly construction barrier to entry. Other provisions allowed regional phone companies to provide long distance service and long distance companies to provide local service after certain performance and penetration levels were reached.

Mergers and other maneuvers have rendered the 1996 Act less effective than was hoped. There is a growing movement to remedy the situation, but as usual, there is little agreement on how to do that—whether it means a revision of the act, a new act, or just a redefinition of the FCC's jurisdiction and regulatory powers.[17]

Different treatment for cell phones, VoIP, and cable TV   The FCC can't regulate the content of cell phone calls or contractual agreements between carriers and customers. However, it does control the portion of the spectrum that cell phones use, can enforce phone number portability,[18] and act on service complaints. In addition, when a cell phone call is routed to or from a landline phone, landline rules apply. Call content remains protected free speech.

The FCC doesn't have direct authority over VoIP because it runs over the Internet,[19] but as with cell phones, it does have some control, especially when calls connect to the landline phone system. Number portability and access to the 911 system are required. Use of particular customer information, such as to whom and when calls are made, is limited. Call content again is off-limits.

As for cable companies, the FCC has some influence over basic service, but by and large, regulation of cable systems is up to state and local governments. Individual communities have *local franchising authorities* (LFAs) that set rates for basic service—local, public access, government, and educational channels—but not for premium channels. Those rates are set solely by the cable companies, although they can't require consumers to purchase anything beyond basic service.

The LFAs can charge franchise fees, which are reflected in service charges. Other fees and taxes find their way onto the bill too. However, not all of the fees are mandated even though their names might imply that they are—some are made up by the cable companies. LFAs also handle customer complaints and can investigate company practices for violations of their franchising agreements.

There also is an important distinction made between those network TV stations that have a broadcast component and those that are strictly cable operations. The former are subject to the same regulations and limitations when carried by cable as when they are sent over the air. Cable stations with no broadcast component are not subject to FCC purview. That's why there is such a disparity in the kind of content some of them carry compared to the broadcast stations.

Venue ownership takes precedence

Are you upset about your treatment at an auto dealership? You can parade on public property outside the business, air your issues over a megaphone, post and blog, walk with signs illustrating your complaint, and your expression is protected as long as you avoid libel or slander. Try to do that inside the dealership. You can be forced to leave and be subject to arrest.

> *Libel* is falsely defaming or damaging a person's reputation in print or by illustration (signs, artwork, etc.). *Slander* is the same thing when the method of delivery is oral. Simply making disparaging remarks in whatever format does not rise to the level of libel or slander unless falsity is involved or reputation is damaged out of hand—often a matter for the courts to interpret.

Is your union protesting what it considers to be unfair hiring practices at a construction site? It can organize a picket line, pose giant inflatable rats, hand out literature explaining the disagreement. As long as the picketing is peaceful, on public space, and not blocking entrances or traffic, free speech holds sway. Move the line to prevent workers or material from entering the site, prevent the flow of pedestrian or vehicular traffic, and the police can shut it down.

Do you wish to expound your views on the sins of a conglomerate, the poor service of a bank, the misleading advertising of a retailer? Blog and post to your heart's content, stand on a soapbox on a public sidewalk and wail away, safe in your exercise of free speech even if your soapbox is in front of the offending institution. Take your protests inside a corporation's offices, bank, or store and your free speech right gets lopped off at the entryway.

The law tells us that free expression in all its forms is protected in public spaces. Owners of non-public spaces can control expression as they will, allowing what they are comfortable with and denying the rest. Of course, they are not completely in control either. So company personnel engaging in discrimination, sexual harassment, defamatory actions, and the like are not protected on free speech grounds whether or not those actions take place on company-owned space.

Protesting the evils of the government at a government site is different because their spaces are a mixed bag. Besides parading in front of government buildings, you can voice or otherwise demonstrate your dissatisfaction inside any of those sites that permit public access. That comes under your powerful First Amendment right to peaceably assemble and to petition the government for redress of grievances. The proviso is wrapped up in "peaceably," though. Disrupting proceedings, blocking access, or otherwise interfering with normal activities except for the distraction your legitimate protest may cause may be allowed for some time but can be stopped if deemed overbearing.

Electronic speech is protected too. You can send emails, blog your opinions, leave voice mail messages, send tweets, and use social media to post your sentiments and organize physical and electronic protests. If you veer off into illegal issues—inciting to violence, libel, slander, and the like, you are subject to lawsuits. Their outcomes, dependent as they are on subjective assessments and interpretations, are not preordained.

Censorship—the one-way street

The philosophy of government in an open democratic society is that its power derives from its citizens and that it exists to protect their rights. As stated in the US Declaration of Independence,

> That to secure these rights, Governments are instituted among Men, deriving their just powers from the consent of the governed.[20]

Accordingly, government suppression of free expression by the public should be invoked only as a last resort. That is quite different from closed regimes, whose philosophy is that government power is supreme, so that securing the authority of and control by their rulers is paramount. Suppression of expression, diametrically opposite free expression, is the first action taken, not the last.

A straightforward ruse from Egypt

Sometimes simple control measures are tried that don't depend on sophisticated technology. During the popular uprisings in Egypt that eventually led to the overthrow of President Hosni Mubarak in 2011, state TV broadcast old images of a deserted Tahrir Square to refute the fact of anti-government demonstrations to Egyptians outside of Cairo. Views of the thousands of protestors that actually filled the square were not aired. Still, a point was reached at which the uprising could not be ignored or made to go away by pretending it didn't exist.

Since constraining free expression by whatever means is vital in closed societies, information flow is managed almost exclusively from the top down. Except for government-sanctioned communication pathways and content, access to information is shut down. A lot of money and effort is expended towards that end. The means of expression that can be sought and used by the general public are strictly limited, as is information flow from the bottom up and among individuals.

On the other hand, information flows liberally from the ruling regimes. They disseminate what they want known to alter reality, to create carefully constructed images of their own societies and distorted ones of those that are at odds with their visions of the world. By keeping their citizenry in the dark as to what life in other societies is really like, leaders aim to make their own images more believable. We're down a rabbit hole, where nothing is what it seems to be,[21] where nothing is what we're told it is.

There is a cat and mouse game going on that has extremely serious ramifications. On one side are regimes that quite effectively use IT to erect barriers to information flow and keep them impervious. On the other are people within and outside those countries who use IT to break through the barriers.

The Chinese experience   China is not the only country that strictly controls communications, but it's a good example of the kinds of actions that autocratic regimes have undertaken. China's leadership has been particularly successful in feeding its population exactly what it wants known, quickly cracking down on dissidents—defined as anyone daring to express a viewpoint different from the party line, challenging authority in any way, or attempting to reach any restricted outside sources of news and other information. This has taken many forms, from erecting techno-logical barriers that block particular sites and means of communication, to thousands of people monitoring the Web, email, and cell phone activities of citizens to ferret out dissidents. All censorship measures were tightened further in reaction to the recent anti-government uprisings in the Middle East.[22]

Internet service providers are a key point in controlling Internet access. China's government requires all connections to be made through its sanctioned ISPs. They, by government order via the State Internet Information Office, will shut down access to particular sites and routes, especially those that lead to international connections. Not all international links are blocked, though. Some are permanently off-limits; others have been selectively blocked and unblocked at different times.

The *Golden Shield Project,* nicknamed by the West as the Great Firewall of China, is a system of firewalls and proxy servers that combine to block access to sites and content based on IP addresses.[23] Ostensibly designed to aid police, it also is used by the government to control access.

The *Green Dam Youth Escort* was software that would automatically prevent access to sites designated as off-bounds. Required to be installed in all computers sold in China, it was touted as protecting China's youth from pornography. Its potential to block any site plus problems it caused with various other commonly installed software resulted in a great pushback from many computer manufacturers and the general public. It was only partially implemented but is still on the drawing board and may make a comeback.

State-owned China Mobile, the largest cell phone carrier in the world, together with Xinhua, China's state-run news agency, is planning to create a new government-controlled search engine. Rather than noting how it will limit information access, it is publicized as a way to "safeguard [China's] information security and push forward the robust, healthy, and orderly development of China's new media industry."[24]

China also engages in a great deal of monitoring of public activity on the Internet. Aside from being able to trace who is attempting to access particular sites by checking site requests at the ISPs and firewalls, many thousands of Chinese are employed to watch Internet activity and screen email content. Screening algorithms have limits and even thousands of people can't watch everyone, so the focus is on known or suspected dissidents. The government determines who is "suspected."

Breaching the barriers

People have become more adept at using social media to get around barriers to information. This is explored in Chapter 6. Here are four popular other means.

- Install a software package that enables *remote operation* of one computer from another. That lets a resident in a closed country connect to a "friendly" computer in another country that has the same software installed and use its unencumbered connections remotely.
- Use an *anonymous proxy server* or series of servers located in other countries to reach otherwise blocked sites. Proxies also can conceal the addresses of the servers that actually fill information requests or post content. Activist groups in many countries provide proxy servers to help those in restricted countries skirt access controls.
- Connect through a *virtual private network* (VPN). Software creates what amounts to a secure tunnel through the Internet to connect to remote networks. The tunnel transports unseen encrypted information between a user at one end and the destination at the other. There are several VPN software packages and charges for the service.
- Install free *Tor* software, which taps into public networks of volunteers outside the country to evade traffic analysis, so that browsing becomes anonymous. That enables information sharing while maintaining communications privacy.[25]

What do you think of this? *Overcoming communication barriers*

Agnieszka Romaszewska-Guzy, a native of Poland, recalled how important it was for Poles to hear news from independent sources—the BBC and Radio Free Europe among others—when Poland was Communist controlled and access to information was very strictly limited. Belarus, Poland's neighbor to the east, was under similar strict control by the Soviet Union (USSR). In 1991, as the USSR was collapsing, Belarus declared its independence. But Alexander Lukashenko, Belarus president since 1994, continued the tight controls characteristic of the Soviet era.

In 2007, Romaszewska-Guzy had an idea. She founded Belsat TV to broadcast news of the world and a variety of other programs out of Warsaw, Poland, reaching into Belarus via satellite. Belsat TV is on air for most of the day. Zerkalo-Info Research Center estimates that it has a Belarusian audience of over three-quarters of a million viewers. Lukashenko considers the broadcasts to be Poland's attempt to destabilize the government of Belarus. Apparently, broadcast television can still pack a punch.

A stealth approach   Under the leadership of the Obama administration, systems are being planned that can reach into restrictive countries to provide direct access to communications that are otherwise unavailable. Internet connections are based on independent wireless networks that can be quickly set up near borders to provide wide area links to the Internet. These access points are designed to be easily moved when necessary. Cell phones are accommodated via independent mobile phone systems that use towers and switching stations located inside protected military bases. Several have been set up in Afghanistan.[26]

Privacy complicates the picture

> *The effort here is for once and for all to give all those who participate in the public process a very, very clear line in terms of exactly what information is protected and what information is not protected.*[27]

As if drawing the line between free and prohibited expression wasn't confusing enough, privacy protection issues considerably confound the situation. The basic problem is that the rights of free expression and privacy can conflict with each other. That dramatically increases the challenge of drawing the line between the precedence of privacy and that of free speech.

As Judge Thomas M. Cooley succinctly put it, privacy is "the right to be let alone."[28] Protection from physical intrusion is one dimension. From an information perspective, it means having control over what is seen, under what circumstances, by whom. That extends easily to cover personal, corporate, and governmental information.

At what point is protection of privacy too much of a throttle on free speech? Conversely, when does exercise of free speech become an invasion of privacy? As usual, in extreme cases it's easy to judge whether particular activities are legal exercises of free expression or overly damaging to privacy rights. The boundary is where the difficulty lies. Different people will have different answers to those issues, as will governments and the courts. There are no absolutes. Drawing the line is a matter of judgment

A fuzzier mandate   Although freedom of speech is explicitly noted in the First Amendment to the US Constitution, there is no such direct guarantee of the right to privacy. In fact, the word privacy is not to be found anywhere in the Constitution or its 27 amendments.[29] There are some oblique references in several amendments that have been interpreted as protecting privacy, however:

- *privacy of belief*—from the First Amendment, not imposing a religion or restricting religious practice, and by extension, choosing not to believe in religion
- *privacy in the home*—from the Third Amendment, not being required to house soldiers, which protects against snooping or discovery by the military
- *personal privacy*—from the Fourth Amendment, prohibition of unreasonable search and seizure, not only regarding property but applied to the person as

well; and from the broadly interpreted *liberty clause* of the Fourteenth Amendment that requires "due process of law" before "depriving any person of life, liberty, or property," which has been cited to prevent meddling in such personal issues as marriage, procreation, raising children, and choosing or denying medical treatment

- *personal information privacy*—from the Fifth Amendment, disallowing mandatory self-incrimination; that is, having the right to refuse to testify against oneself.

Many federal laws have been adopted that deal with privacy issues more or less directly. Here is a representative sample:

> **Two Supreme Court decisions upholding privacy**
>
> *Griswold v. Connecticut* (1965) ruled that Connecticut's prohibition of contraceptive use was unconstitutional on grounds that a decision about whether or not to conceive was a private matter.
>
> *Roe v. Wade* (1973) declared that anti-abortion law in Texas, and by implication such laws in other states, was unconstitutional based on privacy between a woman and her doctor.

Freedom of Information Act (FOIA, 1966)[30] takes the approach that, when it comes to federal government records, the public's right to know supersedes privacy. The reasoning behind the Act, which requires that copies of or access to federal documents be provided on request, is that disclosure keeps the government honest.

Naturally, there are limits to what can be released. Certain records, primarily those declared secret due to defense or foreign policy necessity, identifiable personal information not relevant to document content, law enforcement material for ongoing investigations or trials, and trade secrets, were declared off-limits to FOIA requests.

All 50 states, recognizing the importance of disclosure, have passed their own public records laws that, like FOIA, give the general public access to state and local government records on request. These often are called FOIA requests, though it's the local laws, not federal FOIA, that apply.

As with federal FOIA, certain records are off-limits. Usually fees are involved, which means that information is available only to those with the wherewithal to pay for it. On the other hand, many FOIA and local requests come from news agencies, which then make their findings known to the public.

There also may be considerable delays in fulfilling requests, especially when there are disagreements over whether the records requested are excepted or if they can be released only after particular segments are blacked out. When the records sought are paper documents that must be searched for and copied, delays can be quite lengthy.

- *Privacy Act* (PA, 1974)[31] provides safeguards against invasion of personal privacy through the misuse of records by federal agencies. Among other provisions, it defines FOIA limits on disclosure of personal information more clearly and requires correction of erroneous information on request.

- *Right to Financial Privacy Act* (RFPA, 1978) protects confidentiality of personal financial records as do parts of the *Gramm–Leach–Bliley Act* (GLB, 1999).[32]
- *Electronic Communications Privacy Act* (ECPA,1986)[33] safeguards information from interception while being transmitted electronically.
- *Health Insurance Portability and Accountability Act* (HIPAA, 1996)[34] shields personally identifiable health information from being revealed without consent.

An important piece of legislation, called the Communications Decency Act of 1996, which is actually *Section 230* of Title V of the Telecommunications Act of 1996, has contributed greatly to the ability of ISPs to provide all sorts of information. It states in part:

> No provider or user of an interactive computer service shall be treated as the publisher or speaker of any information provided by another information content provider.[35]

This proviso protects online services and users of those services from liability based on content that was created by someone else, regardless of whether the target of the content considers it to be invasion of privacy, offensive, defamatory, reputationally harmful, or even false. The services also are free from liability if they prevent access to particular material or provide others the means to do so.

Section 230 does not protect the original content creators. They can be held liable for disparaging material, copyright violation, and other abuses. But the Section is controversial because aggrieved parties have nowhere to turn for relief if they can't find the original creators or if they are in countries where laws regarding libelous or invasive actions are non-existent or unenforced.

Taking a less formal approach than legislation, the Federal Trade Commission (FTC) developed a strategy released as a report called *Fair Information Practice Principles* (FIPP, 1998).[36] A comprehensive model for information collection and use by online entities, it details what the FTC considers to be appropriate activity and practice. The report comprises guidelines, not laws, though they have been generally accepted in principle if not always practiced sincerely and candidly.

The first five sections focus on principles of practice in specific problem areas, the last two on applying those principles to children's issues. Quoting excerpts from the first five by way of summary:

1. *Notice/Awareness*
   The most fundamental principle is notice. Consumers should be given notice of an entity's information practices before any personal information is collected from them. Without notice, a consumer cannot make an informed decision as to whether and to what extent to disclose personal information.
2. *Choice/Consent*
   At its simplest, choice means giving consumers options as to how any personal information collected from them may be used. Specifically, choice relates to secondary uses of information—i.e., uses beyond those necessary to complete the contemplated transaction.

3. *Access/Participation*

. . . an individual's ability both to access data about him or herself—i.e., to view the data in an entity's files—and to contest that data's accuracy and completeness.

4. *Integrity/Security*

To assure data integrity, collectors must take reasonable steps, such as using only reputable sources of data and cross-referencing data against multiple sources, providing consumer access to data, and destroying untimely data or converting it to anonymous form.

5. *Enforcement/Redress*

. . . the core principles of privacy protection can only be effective if there is a mechanism in place to enforce them. Absent [that], a fair information practice code is merely suggestive . . ., and does not ensure compliance with core fair information practice principles. Among the alternative enforcement approaches are industry self-regulation; legislation [providing] remedies for consumers; and/or regulatory schemes enforceable through civil and criminal sanctions.

From a common-sense viewpoint, the Principles are logical and reasonable. So after more than 13 years, how have they been applied? In a word, loosely. The record is quite spotty. Even where the guide is followed, it often is more a matter of technical compliance than meeting the intent of the principles. Practice leaves much to be desired.

Regarding Notice/Awareness and Choice/Consent, many organizations do not make obvious what they collect and what they do with the information. Others outline their practices as densely written overly long terms of use to be agreed to before proceeding on a Web site. Still others provide notices of privacy practice with instructions on how to make choices, but typically they also are densely written and overly long, with little privacy protection in the absence of specific requests to the contrary.

As for Integrity/Security, it is a rare site that allows us to access our own data, much less to correct errors. The three major credit reporting agencies, Trans Union, Equifax, and Experian are exceptions.

Enforcement/Redress is the least followed of the guides. Industry self-regulation is weak and laws protecting privacy have not kept pace with advances in IT.

Disclosure   Every country has laws and practices supporting and contravening privacy. They cover the gamut from the European bent of heavily favoring privacy rights to the rigidly constraining practices of autocratic regimes where any sort of privacy is far from guaranteed. In the US, FOIA and FOIA-like requests put the people's right to know ahead of government privacy, but individual privacy isn't always sacrosanct either. Personal information disclosure takes precedence over privacy in a great many everyday activities.

For example, we have to provide personal information to: establish a bank account, apply for a loan, and qualify for a credit card; take out an insurance policy; apply for financing from an auto dealership or manufacturer; verify and satisfy obligations to a tax authority; fill out an employment application; comply with federal enforcement agencies and the police during investigations; serve the courts for jury selection and trials. At least we know that we are supplying personal information in those cases, by choice or not, but we also give up privacy in subtler ways, discussed in the IT affects outcomes section below.

What about the press? Being specifically mentioned in the First Amendment, the press enjoys the same free speech rights as people. That means the press isn't constrained from publishing information it comes across, even if it may seem to cross the privacy line. Of course, laws relating to libel and copyright infringement apply.

Confidentiality of sources, even when the information was illegally obtained, has been a line in the sand that the press crosses only in the most extreme cases. Their justification for source anonymity is that without it people would be less likely to provide important information of interest to the general public since it might put them at risk of reprisals or physical danger if their identities were known. Identity protection is so strongly held that reporters have gone to jail in contempt of court for refusing to reveal their sources.

Reputable publishers take internally mandated reasonable, though not foolproof, steps to assure information is legitimate. Judgments often include considering whether publication is wise even if all appropriate conditions are met. Where mistakes have been made, corrections are usually published. If harm is done, lawsuits and settlements may follow, just as when individuals are harmed.

What do you think of this? *The credit check phenomenon*

- Want to take that car for a test drive? Sure, just let us make a copy of your driver's license. While you're out on the road, the dealership runs a credit check on you so that if you get to the bargaining stage, they know what risk category you are in.
- Get an unsolicited credit card offer in the mail? That's likely the result of a credit check by the card issuer who likes your score and wants you as a cardholder.
- Applying for a job? Many employers run credit checks as a factor in deciding how responsible, trustworthy, and truthful you are. A bad rating could mean no job offer.
- Want to rent an apartment? The landlord will likely run a credit check to gauge your ability to pay the rent.

These are the tip of the iceberg. Credit checks, explicitly authorized by you or not, reasonable or not, are becoming more widely used in all sorts of situations. Moreover, it's often not clear or made apparent how information from those checks is evaluated in reaching judgments, especially those not directly related to financial issues.

Can a leopard change its spots? Can you be forever marked by unwise financial choices? A bad credit history can follow you for a long time in the minds of the evaluators, even when the recent past indicates that you've mended your ways.

Public lives, private lives   Politicians, celebrities, and sports stars are in their own universe. They use IT to further their careers and influence, crafting public personas by generating as much buzz as they can. Far more than a one-way street, the IT of the day—Twitter, Facebook, email, and interactive Web sites—draws in the public as participants in the exercise. In doing so, they trade away a large measure of their privacy.

That may be all to the good, but technology is a neutral player, enabling the beneficial and the damaging with equanimity. And so when unflattering or disparaging information is online, it can be as ruinous to a reputation as the good buzz is a boost, and often more quickly. Here is one recent example:

> A married congressman carries on virtual affairs via sexually explicit emails, text messages, and tweets. Despite denials, obfuscation, and efforts to contain the information, it comes to light. He is publicly castigated, resigns under pressure, and faces the end of a promising political career.

In a sense we, as "ordinary" citizens, are in the same situation. When we go online, some of our private lives becomes public too. Privacy boundaries are broken when evidence obtained from online communications is used in divorce cases, child custody trials, property disagreements, civil disputes, and criminal indictments. For most of us, the good and bad outcomes are local—after all, there's not much about us that interests the world at large. But in that narrow way, we run the same risks as those in the limelight and on a personal level they can be just as harmful.

So where to draw the line between public and private? How far can responses to bad judgment go and still be considered justifiable? Let's look at a few examples. You be the judge:

- A highly qualified job applicant with a stellar résumé isn't hired because a search revealed Facebook photos of him at a party looking somewhat drunk and holding a beer in each hand. Is the hiring denial a legitimate response? An invasion of privacy? A violation of the applicant's free speech rights?
- Suppose that applicant was hired without knowledge of the photos, but subsequently they turned up and he was fired—what then?
- A firefighter posed for revealing photos while wearing her fire station hat and only some of her firefighting gear, holding a hose in a suggestive pose. The pictures were posted and came to the attention of the fire commissioner. She was disciplined, demoted, and put on a non-paid leave for a year. Justified?
- What if the same sort of pictures were posted except that nothing identified her as a firefighter, although it was readily obvious to her coworkers and supervisors, who would have no trouble recognizing her?
- A high school teacher, having dealt with a particularly disruptive class all semester, tweets that he'd like to strangle some of the boys in the class, a remark he never acted on. His comments were picked up and circulated, outraging parents. The teacher was fired. Were his rights infringed?
- An employee complained about the difficulty of working for her boss. In her frustration she vented by posting descriptions of particularly annoying incidents

on her blog. The boss caught wind of her postings and she was passed over for promotion, though she was the most qualified. Acceptable?

In each of these situations, we could say that when personal peccadilloes become public knowledge, they reflect on the organization in question and so are rightly part of personnel decisions. After all, what we have here is just a matter of proper use of technology to help in assessing employee suitability, a step up from résumé review and background checks.

On the other hand, we could say that a line has been crossed. We are entitled to our private lives. They are no one else's business and should not enter into workplace decisions. If the person is doing the job, that's all that counts.

IT affects outcomes

Before the Internet, it's quite likely that none of these situations would have come to light. The job applicant and the firefighter might have put their photos in albums, stuck them on refrigerators, or stored them in shoeboxes, but the HR department and the fire commissioner would not likely have seen them. The teacher might have complained to his family or friends, but his frustrated outburst probably wouldn't have been heard by anyone else. And without a blog platform, who but a few selected friends and family of the employee with the difficult boss might have ever heard of her aggravations?

When business and government records were solely paper documents stored in vast arrays of file cabinets, you would have a difficult and tedious time invading privacy. Once you gained physical access, legally or otherwise, what could you do to take away what you found—carry away drawers full of folders, photograph or Xerox page after page, copy records by hand? And once you did, you could distribute the information widely only by bringing it to a newspaper, a radio or TV station and hope that they would publicize it.

The digital era and the Internet changed all that. With growing frequency, records are captured digitally and held in databases. Massive projects to digitize print and other analog information are underway as well. Getting access to a digital database, legally or otherwise, is all that it takes to download every record it contains quickly and remotely. Technologically speaking, if it's connected it's reachable one way or another.

We trade privacy for convenience in many ways. The issue is, do we know when and how we are doing that? Sometimes we do. Often we don't.

Is your supermarket customer loyalty card scanned? Your food preferences are revealed. Do you use a transit fare card or an RFID device[37] to pay highway tolls? Your trips and times are mapped. Are your paychecks deposited directly to your bank? Your salary history is recorded. Do you pay bills online? Your utility use, favorite stores, and overall shopping activities, with dates and times, are uncovered. Do you carry out searches online? Your quests are profiled. Do you download material to an e-reader? Your reading habits become clear. Do you use a credit card? Your expenditure patterns are an open book.

Most of us are aware of these practices, although not always what's done with all those data. It turns out that information, our information, is a valuable commodity that is routinely bought and sold. Presumably our privacy is preserved by removing personally identifiable information from the data. Rather chillingly, some analysis techniques have been demonstrated that pierce the anonymity veil.

Identifying the unidentified   Information that has been anonymized by removing such personally identifying data as names, social security numbers, and addresses can often be de-anonymized to reveal the person behind the data. The most common technique, though not the only one, combines two data sets—the anonymized data and one with unrelated but generally available data, such as voter registration rolls and census reports, that identify individuals. By linking the two sets on a common field that's usually not removed from anonymized data, such as zip code or birth date, the identity of the anonymous individuals can be discovered.

- In 1990, Professor Latanya Sweeney[38] found that zip code, birth date, and sex from data sets that had names, addresses, and social security numbers removed could be combined with census data to uniquely identify 87 percent of the United States population.
- In 2006, Netflix supplied researchers with the movie ratings of 500,000 members, with names removed. They were able to identify 99 percent of the individuals in that set using just six reviews of each of them.
- In a 2009 study, Professor Alessandro Acquisti[39] and researcher Ralph Gross[40] showed a two-step process that used birth data and geographic location to predict an individual's social security number.[41]

With our knowledge and without   It's one thing when we're aware of the tradeoff we make between convenience and privacy. It's another when our privacy is breached unbeknownst to us. Two significant examples are location tracking and cookies.

Location tracking   Whenever our cell phones are on, whether or not we are using them, they regularly communicate with the base stations of the cells we are in. A record of our locations and movements, no longer private, is stored even though we are not making or receiving calls and are not active participants in the process. Smartphones almost universally have GPS service, which makes tracking more precise and is operative whether or not we are seeking route guidance.

Carriers also keep call records—not call content but who the calling and called parties are, their locations, and the time and length of the calls. These records can be used by law enforcement to identify suspects, find them, make arrests, and as trial evidence. It's hard to hide anymore, which may be a good thing but for sure is another chip off our privacy block.

Marketers are particularly interested in location data. When the cell phone is a smartphone, location information can be used to send ads and notices for nearby retail establishments. A convenience, possibly, but the carriers collect fees for that service and store that information as well.

More specific information is gathered when we deliberately make use of location-based social networking applications like Foursquare, Gowalla, and MyTownNavigator to signal our whereabouts or find deals. No secrets there, but that information is also used for directed ads and click-throughs for which the application providers get paid. And, of course, records of our travels, searches, and purchases stay in their databases whether we continue to use those apps or not.

Cookies   A cookie is a very small text file that a Web site creates and saves on your computer the first time you visit the site. It contains tags, updated on subsequent visits, that relate to information stored on the site's servers about you and your activities on the site. For example, sites that offer items for sale commonly collect name, billing and shipping addresses, credit card numbers, other payment methods, contents of shopping carts and other lists, items searched for, those looked at more closely, items bought, ratings, opinions, likes and dislikes. The vast majority of Web sites use cookies.

Cookies save us a lot of time and, as site owners like to say, enable them to provide a better user experience. When you return to the site, you may be greeted with suggestions based on your past searches and purchases. Your shopping cart and wish list are readily available. If you purchase something, your shipping and payment information load automatically. Some sites, especially Web portals, allow you to customize screens by specifying what you want shown and where, that information being stored on the site linked to via cookies. That's the good news.

What is done with that information? One profitable use is for targeting, not only for suggestions by the site itself but from advertisers. Companies pay more for ads that relate closely to user profiles that a site assembles from the data they collect. Those ads can reach us in pop ups, drop downs, and fixed places on the site, by email, and on our smartphones—helpful perhaps, but maybe a lot more than we want. Wouldn't it be nice to escape from ad bombardment once in a while?

Targeted ads may be off-target, hence even more annoying. They assume our profiles are accurate. But if we search for something of only passing interest, or that we want to keep current on but have no intention of purchasing, or that we purchase as a gift of no relevance to us, that becomes part of our profiles too, shifting the target.

What do cookies have to do with privacy? Ostensibly, only the site we visit can create a cookie and only that site can access it. That would seem to make privacy less of a concern. But there is a clever and controversial way around that restriction that takes advantage of the fact that, rather than handling ads that appear on their sites, most contract with media service companies to supply the ads for them.

When a Web page opens, a number of commands are sent by the browser to fill in the different parts of the page. If ads are to be placed, commands go to the media service, which sends the ad images and, importantly, can also set cookies even though we haven't visited the media service's site. Then the media service can use those cookies to capture information about our activity on the site just as the site's cookie can. If the same media service is used to place ads on other sites we visit, it can combine the information to create a much fuller and more detailed profile of us that advertisers will pay more for.[42]

We can set our browsers to reject cookies, but that usually blocks us from many features of the sites and in some cases from entering the sites at all. We can delete a cookie after we close a site, but that negates the convenience advantages that cookies bring to our next visit. Moreover, the information that was collected remains stored at the site.[43]

*   *   *

These two information technologies are examples of the consequences of our choosing convenience over privacy. What we buy, what we eat, where we travel and by what means, the channels we watch on TV, the movies and videos we stream, our income and expenses, where we live, where we work, and much more is already being collected and stored. We are rapidly moving closer to the time when just about everything there is to know about us is captured and saved somewhere.

Personalization revisited   We looked at some personalization issues in Chapter 2. Here we can make an interesting link between privacy, personalization and de-anonymization. In the business world, personalization is widely used to create profiles that characterize us in various ways—by what we buy, search for, like and dislike, our incomes and their sources, where we live and in what kind of homes, our food preferences, marital status, vacation choices—everything known about us. Combine personalization information with de-anonymizing techniques and privacy goes out the window.

Personalization also is the branch from which hangs the information we are most likely to see, the basis for targeted ads and directed searches. And that's both the positive and negative side of the story—positive because we see more of what we want to see; negative because we miss seeing other views of the world.

Personalization's analog, data profiling   In much the same way as personalization profiles us, data profiling characterizes businesses and their competitors. Profiles are created by delving deeply into the mountains of data that businesses collect and store, resulting in summaries and drilled down details that are used to make strategic decisions and to uncover anomalies that need attention. That's the good news. What about the tremendous stores of data that businesses have about us that can be mined to extract information that we might prefer to keep private? That's an area where we have no control.

Businesses may say that in their analyses they have no interest in personal identification, but a lot of information about a business and its competitors is personal, hence identifiable. So aside from their claims, what evidence do we have that our privacy is protected?

What do you think of this? *May I have my own data, please?*

Nigel Shadbolt, professor at the University of Southampton, England, believes that "Information about how we shop, travel, communicate and live our lives is a powerful source of insight, so it's important that each of us should be able to access our own data." He has been appointed chair of the *mydata* group, part of an initiative launched by the British government on April 13, 2011, called *Empowerment Strategy; Better Choices: Better Deals. Consumers Powering Growth* (http://www.bis.gov.uk/assets/biscore/con-sumer-issues/docs/b/11-749-better-choices-better-deals-consumers-powering-growth.pdf).

The idea is to see what government can do to improve consumer power by giving them specific knowledge of and simple access to the data that companies collect about them, as well as the ability to download their data and safely pass them on to others if they so desire. One of the results of this empowerment is expected to be a boost to economic growth because of the increased influence consumers will have. *Mydata* expects to develop non-regulatory partnerships with businesses and consumer organizations.

(*Information World View*: http://www.iwr.co.uk/news-and-reference/
3010872/Nigel-Shadbolt-appointed-chair-of-mydata-project)

The whistleblower—risk and reward

A whistleblower is a person who publicizes illegal or deceitful activity committed by those in an organization or is sanctioned by an organization—government agency, business, or any public or private institution. The uncovered activity may be presented to persons of authority in the organization or revealed to the public at large. Typically, whistleblowers work for the organization whose misdeeds they've discovered, which is how they find out what's going on. And also typically, their identities are known.

It would seem that whistleblowers are a noble lot, playing an important role in keeping organizations honest, and in many instances that's what they are. Of course, not every revelation is monumental. Many are just simple issues that are handled internally without further ado. When more serious matters are discovered, their disclosures are not without controversy or risks to the whistleblowers themselves as well as to the organizations.

Whistleblowers are both hailed as heroes exercising free speech rights and reviled as invaders of privacy betraying privileged information. They are both portrayed as being truly disturbed by what they find and as simply seeking money and renown.

However they are viewed, they are at some risk when they take the step of revealing what they've found. Protection is spotty in coverage and application, with a patchwork of federal and state laws that sometimes are contradictory. Whistleblowers have been fired, demoted, and otherwise forced to leave. Even when in the right, they may bear the mark of untrustworthy employees and so, by coming forward, run the risk of jeopardizing their futures and careers.

Laws and protections, more or less   There are two kinds of whistleblower laws—those that provide procedures for whistleblowers to act effectively and those that protect whistleblowers from reprisals for their actions. The most widespread of the former invoke *qui tam* law, which enables a whistleblower to file suit against any person or organization involved in improper activity involving federal or state governments.

Qui tam cases concern covered up violations of legislative acts, misreporting, falsified data or evidence, and disclosure failure. Governments often participate as litigants with the whistleblower because qui tam suits can be considered as being filed on their behalf.

The major qui tam federal act that subsequent laws have built on was passed in 1863 in reaction to fraud by suppliers to the Union military during the Civil War. Called the *False Claims Act* (FCA), colloquially known as the Lincoln Law, it included a qui tam provision that also allowed for a percentage of any funds recovered to be paid to the whistleblower. The provision continued under the 1986 revisions to the Act,[44] which made suits easier to pursue and increased potential awards to the whistleblower. That's what leads aggrieved parties to complain that the true motive of the whistleblower is financial gain, which in cases of large-scale fraud can amount to a lot of money.

The federal *Fraud Enforcement and Recovery Act* (FERA),[45] passed in 2009, expanded the provisions of the 1986 FCA, clarifying and liberalizing what constitute proper claims. Several other more or less related laws have been passed as well, aimed at particular industries. For example, the *Dodd–Frank Wall Street Reform and Consumer Protection Act* (WSRCPA),[46] passed in 2010, gives the Securities and Exchange Commission authority to reward those exposing fraud at public companies.

As for whistleblower protection, the major federal provision comes from the *Occupational Safety and Health Act* (OSH Act, 1970, amended 2004),[47] administered by the Occupational Safety and Health Administration (OSHA). The Act's goal is to ensure safe and healthful workplace conditions. To reach that goal it depends to a large extent on whistleblowers—employees relating instances of contrary practices, asking for an OSHA inspection, and testifying about activities in violation of the Act.

Significantly, the Act protects employees against retaliation for their revelations. Moreover, it prohibits employers from creating a climate that discourages employees from taking appropriate revelatory actions. Seventeen other statutes that deal with whistleblowing in specific industries also are administered by OSHA.

WikiLeaks—global whistleblower, global hero, global villain

> The digital age has changed the dynamics of disobedience in at least one respect. It used to be that someone who wanted to cheat on his vow of secrecy had to work at it. Daniel Ellsberg tried for a year to make the Pentagon Papers public. There was a lot of time to have second thoughts or to get caught. It is now at least theoretically possible for a whistle-blower or a traitor to act almost immediately and anonymously.[48]

In their own words,

> WikiLeaks is a non-profit media organization dedicated to bringing important news and information to the public. We provide an innovative, secure and anonymous way for independent sources around the world to leak information to our journalists. We publish material of ethical, political and historical significance while keeping the identity of our sources anonymous, thus providing a universal way for the revealing of suppressed and censored injustices.[49]

WikiLeaks qualifies as an "interactive computer service" provider under Section 230 of the Communications Decency Act, and therefore is protected in the US from liability for the content it distributes that was created by others. So it would seem to be a no-brainer that when WikiLeaks came into possession of many thousands of classified US government and military documents and over 250,000 secret embassy cables, it would be seen as acting within the law if it published any of them.

That was not what happened. Instead, WikiLeaks was strongly pressured by the governments of many nations and their intelligence agencies across the globe to persuade the organization to not publish the documents and to delete them from their servers. Most of this was to no avail. And so WikiLeaks founder Julian Assange was pursued in an attempt to discredit him and, by association, WikiLeaks. He was accused of rape and sexual molestation by two women in Sweden. An international arrest warrant was issued by Sweden seeking his extradition so he could be brought to trial there.

Also sought were the sources of the information WikiLeaks obtained. That person or persons would be fair game under the law since the classified and secret documents were illegally obtained and supplied, but WikiLeaks promises anonymity to content suppliers just as does the traditional press.

So we come to a key question: Is WikiLeaks providing a vital legitimate service that keeps governments honest or is it jeopardizing nations and their alliances by releasing confidential documents detailing internal workings of those governments? Reactions to the leaks were wildly different. Here are three examples from the US:

- Representative Peter King (Republican, New York), ranking member of the House Homeland Security Committee, focused on WikiLeaks itself, saying that WikiLeaks should be classified as a foreign terrorist organization and that Assange and the organization should be prosecuted under the Espionage Act. Then the US could "seize their funds and go after anyone who provides them with any help or contributions or assistance whatsoever."[50]
- Representative Pete Hoekstra (Republican, Michigan), ranking member of the House Intelligence Committee, turned his criticism on the government, calling the leaks "a massive failure within the intelligence community to create this kind of a data base with this much information in it and then provide access to it to hundreds of thousands of people across the government."[51]
- Senators John Ensign (Republican, Nevada), Joe Lieberman (Independent, Connecticut), and Scott Brown (Republican, Massachusetts) introduced legislation that added to the provisions of the Espionage Act. Called the *Shield Act*,[52] it makes it illegal to publish the names of informants of the US intelligence

community and military. Presumably that would give the government flexibility in going after Assange, except he's not a US citizen and so is not directly indictable by the US.

Security—yet another complication

We want to be secure in our being, free from intrusions surreptitious or overt, confident that our privacy is protected. At the same time, we want to be able to seek information and express our opinions without restrictions, fear of reprisal or retribution.

We've seen potential conflicts between free expression and privacy. Another step up in complexity comes when security enters the mix. Security can conflict with both free expression and privacy—when security interests override privacy concerns, when security is threatened by free expression.

Overall, reactions to the WikiLeaks revelations are of the "locking the barn door after the horses have escaped" variety. The real issue is how security of sensitive documents is managed in the first place. Apparently, not very well—too many people have access to too many documents that they have no real need to see. That's one area to focus on to prevent future leaks. Others include strong encryption, more rigorous security and need-to-know clearances, and blocking interconnections among particular types of information and particular storage resources.

These and similar actions would be helped considerably by more prudently deciding which documents really need to be classified as secret at all, or at the various higher levels of secrecy, and how long they need to be kept secret before being declassified. The fewer actually secret documents there are, the easier it is to protect them.

We undertake a more extensive discussion of security issues in Chapter 7.

And so—where do we stand?

We've arrived at a point where privacy is, in many respects, incompatible with online and mobile communications. We cannot be assured that what we believe to be secret does not become public. When we take to the convenience of the Internet or use cellular networks, to a great extent we leave privacy behind. The same is true of every organization—businesses and governments alike.

We know that we must take responsibility for what we purposely make public and what can easily become public—what we post, tweet, and blog, the messages we send and the emails we write. After all, if it's not online, the chances that it won't be revealed are much improved. So yes, we do need to be mindful of our activities, but privacy incursions can be beyond our diligence and control and, as we've seen, can take many forms. Importantly, can disparities be reconciled?

IT on the job   We like to think of information technology as a neutral enabler, neither inherently good nor evil. In practice, it is used effectively for beneficial purposes and for destructive ones. The Internet and wireless devices intensify the possibilities of both, though the ones that make the biggest news are usually the most negative. Some examples from around the world:

- Rules issued by India's Department of Information Technology in 2011 require a Web site to take down content that citizens and officials find objectionable, as outlined in a long list that includes "threatens the unity, integrity, defense, security or sovereignty of India, friendly relations with foreign states or public order."[53] Prohibited is information that "is grossly harmful, harassing, blasphemous defamatory, obscene, pornographic, paedophilic (*sic*), libellous (*sic*), invasive of another's privacy, hateful, or racially, ethnically objectionable, disparaging, relating or encouraging money laundering or gambling, or otherwise unlawful in any manner whatever."[54] These terms are nebulous enough as to be interpretable to suit and applicable in a tremendous variety of situations, making free speech in India an endangered species.
- In May, 2011, US Citizen Lerpong Wichaikhammat was arrested in Thailand, accused of insulting Thailand's King Bhumibiol Adulyadej on the Internet. Thailand's Department of Special Investigation said he translated and posted on his blog an article considered offensive to the king, and a link to a banned biography of the king. Thailand's laws prohibit insulting, defaming, or threatening anyone in the monarchy. These prohibitions are quite vague, easily applied at will by the monarchy and those assigned to protect it.

What do you think of this? *The inquisitors*

The name of the Office of Prevention and Suppression of Information Technology Crimes, established by Thailand's constitutional monarchy, implies a government effort to curb criminal activity supported by IT. The question is, what is an IT crime in Thailand? The answer is, a crime is anything deemed offensive to the king or his family. The monarchy justifies this stance by claiming that it is needed to protect the morals of the populace and promote peace, but the prohibitions are so ambiguous as to be interpretable to suit the whims of the Office, and presumably those of the monarchy as well.

As a result, even the vaguest "suspect" remarks that can be attributed to a person or group are rarely attempted. Is this an illustration of the value of preserving anonymity on the Internet?

- Considering how popular uprisings have affected autocrats in neighboring countries, Iran set up a *Facebook Infiltration Task Force* whose job is to scour the site for anything considered to be inimical to the ruling regime and to block citizen access to pages with such content. What is inimical is broadly interpreted and applied at the whim of the Task Force and the regime's rulers. Punishments for claimed violations can be harsh.

- Reacting to citizen unrest and the possibility of ever larger demonstrations against the monarchy, Syria shut down access to YouTube and Facebook in May, 2011. The *Syrian Revolution 2011* page on Facebook, with over 180,000 members, was a particular target. In a twist, Facebook closed down the 80,000-member *Syrian Electronic Army* page aimed at disrupting protestor moves, because it showed how to conduct online attacks against opponents, which violated Facebook's terms of service.

- China once again demonstrated its drive to swiftly punish free speech when it remanded Hong Fang to a year in a Chongqing re-education-through-labor camp. A retired forestry worker, he committed the offense of criticizing Chongqing Communist Party leader Bo Xilai by posting an online verse that in addition to the critique used rather crude language. It seems likely, though not completely clear, that in the absence of the latter, the same punishment would have ensued.

- In a bizarre quirk of law, lawyers defending Guantánamo detainees are forbidden to read online any of the Guantánamo related documents released by WikiLeaks. The US State Department declared that even though made public, the material is still classified information and must be treated as such. That means lawyers may have access only after proper clearance and then only in particular secured locations. Even more bizarrely, government employees, military personnel, and their families who view the documents online could be prosecuted under the terms of the Espionage Act. Members of the general public could lose any chance of qualifying for government jobs. This is a striking example of the law lagging behind the technology.

Which leaves us where? The Internet and mobile communications networks greatly enhance opportunities for free expression, so access for all is considered to be a worthy goal, though one that must be approached with caution. As is stands, the Internet itself is an inherently porous system. So are wireless networks. By hook or by crook, anything connected to the Internet is reachable and anything sent over the airwaves can be intercepted.

We must remember that it takes continuous vigilance to keep the villains at the gate. When we're careless with our online and communications behavior, we leave the gate wide open. This applies at least as much to businesses and government organizations as to us. Too many breaches of what we believed to be secured information have resulted in too much private data being released.

So here are important questions to think about:

- Do we trade away privacy willingly, carelessly, or unknowingly?
- Should there be more transparency about what is collected, by whom, and under what circumstances?
- Should it be easier for us to control what is gathered and what uses it can be put to?
- Should collected information expire after some suitably short time after which it must be deleted?

- And what about closed societies—is it our responsibility to do what we can to help their populaces overcome information access barriers or is it best left to the people of those societies themselves?

It might be nice to believe that we will reach the day when relatively unfettered global access is a fact rather than a dream, but think about how much more of the world's information will be at peril as that comes closer to reality. The more connectivity there is, the more information that will be collected and sought, and the more at risk we, organizations, and governments will be for misuse, abuse, and release of sensitive information.

We are moving farther away from privacy and control of our personal information. If this trend continues we may be headed for a point where for all practical purposes privacy is a thing of the past. If we want to save privacy from being relegated to an outmoded concept, the law is a good place to start. But it is not a foolproof place.

Although privacy concerns have been voiced for some time, there still is no general law dealing with all aspects of privacy. That may not be a bad thing, though, since excepting a miraculous outcome, it's likely that crafting a truly comprehensive law would result in an opus too big to comprehend and too convoluted to be useful. Then too, we've seen repeatedly how quickly technology renders legislation obsolete, or at least reduced in power and influence. Even so, law-based societies must look to the laws and keep them as relevant as possible.

On the bright side, there is legislative movement on particular privacy fronts that deserves watching and even cheering on. To a large extent, this has come about as a result of consumers and privacy rights organizations pushing back, with focused efforts that are aimed at giving privacy a fighting chance.

One of the latest important attempts came in the form of a bill proposed by Senators John Kerry (Democrat, Massachusetts) and John McCain (Republican, Arizona) called the *Commercial Privacy Bill of Rights Act* (CPBRA) introduced in April 2011.[55] It has created a stir, both for what it provides and what it doesn't. At this juncture, it may be that half a loaf is better than none.

On the privacy protection side:

- Businesses with Web sites would be able to collect only the information needed to provide specific services to their users and customers and they would have to strengthen security to protect data from unauthorized access and misuse.
- Businesses would be allowed to market to any consumers, but consumers would be able to opt out of receiving behavioral (targeted) ads and keep information from being transferred to or used by third parties.
- Use of sensitive personal information would require consumer opt in. Furthermore, users would be able to see collected information and correct it or have it removed.

Objectors point to several shortcomings and loopholes perceived as watering down the bill's effectiveness.

- A "do not track" option, such as the one called for by the Federal Trade Commission, is missing. That would have prevented companies from following activities on the Web for those who choose the option and would have required browsers to provide that choice.
- Although legal actions against violators of the bill could be brought by the FTC or state attorneys general, they could not be brought by private individuals even if the violations caused personal harm.
- Companies would be allowed to implement their own privacy policies if they are judged to be compliant with the bill by the Department of Commerce, whose main thrust is supporting businesses, rather than by the more consumer-oriented FTC.
- And social media marketers would be exempt from some of the information collection protections.[56]

Another effort is the *Location Privacy Protection Act* (LPPA) proposed in June 2011 by Senators Al Franken (Democrat, Minnesota) and Richard Blumenthal (Democrat, Connecticut) that would require obtaining permission from subscribers before information collected by mobile service providers could be shared with third parties.[57] The bill is intended to close loopholes in various federal laws that leave users vulnerable to having their information shared without their knowledge and without the ability to prevent that from happening. Privacy protection requirements are strengthened and operators would have to delete information if requested to do so by customers.

In the same month, inspired by complaints over Facebook's facial recognition technology, the *Electronic Privacy Information Center* (EPIC) filed a complaint with the FTC in June, 2011, asking that Facebook be barred from using the technology.[58]

A different approach was taken by Senator Patrick Leahy (Democrat, Vermont). One of the authors of the 1986 Electronic Communications Privacy Act (ECPA), he realized it was in desperate need of updating, having been formulated before most of today's privacy and security issues existed. He proposed the *ECPA Amendments Act of 2011*[59] that would require government agencies to acquire search warrants before accessing stored electronic communications information or stored location data. Protection is not complete, however. Warrants are required only for records no more than 180 days old. Otherwise, searches can proceed with just a prosecutorial subpoena that can be issued without informing the person involved. Further, agencies have 90 days to notify a person that they've had their records searched.

Governments of all nations have been pushing back too, seeing access as a two-edged sword—one edge that can be used to sway populations, the other that can put administrations at risk of loss of power. Even in autocratic countries where long tradition has kept control consolidated in the hands of supreme rulers, the rising tide of information is lapping at the seats of power. This is looked at more closely in Chapter 6.

So we've seen that drawing a line between what's acceptable and what's not can be difficult, whether concerned with free expression or privacy. We've also seen how the task is complicated when emphasizing one of the two comes at the expense of the other. As IT inexorably advances, we can expect the tug of war within each area and between the two to continue.

The lines that strike the balances are continually being examined and repositioned. The lesson is that determinations must always be for the moment. They cannot be set in stone, but will and must shift as the needs and desires of societies adjust and transform.

Notes

1 James Madison (1751–1836), fourth president of the US from 1809 to 1817: http://www.brainyquote.com/quotes/quotes/j/jamesmadis392906.html
2 Nicolò Perotti, writing in 1471 in reaction to the impact of Johannes Gutenberg's printing press invention, circa 1440.
3 The Confederation Congress succeeded two Continental Congresses. Altogether, from 1774 to 1789, they were the legislative body of the Confederation of States, which became the United States after ratification of the Constitution.
4 James Madison has been called the father of the Constitution (see note 1).
5 Actually titled *The unanimous Declaration of the thirteen united States of America*, the full text of the Declaration of Independence is at: http://www.unalienable.com/transdec.htm
6 Ira P. Robbins, Professor of Law and Justice, American University, Washington College of Law, as quoted in *The Oregonian*'s Web site: http://www.oregonlive.com/clackamascounty/index.ssf/2010/02/clackamas_man_exercises_free_s.html
7 Voltaire, the penname of François-Marie Arouet (1694–1778), was a philosopher and writer during the Age of Enlightenment, a period when reason was considered to be the basis for authenticity and influence.
8 For the full text of the law, see: http://law.justia.com/codes/california/2009/civ/1746–1746.5.html
9 For the full text of the opinion, see: http://www.supremecourt.gov/opinions/10pdf/08–1448.pdf
10 Spencer Tunick (1967–). Two cases in point in 2004: Cleveland, Ohio, June, 2,754 people posed nude; Buffalo, New York, August, 1,800 people posed nude. He has done similar installations in many countries around the world.
11 For the full text of the 1934 Act, see: http://www.criminalgovernment.com/docs/61StatL101/ComAct34.html
12 The electromagnetic spectrum is a scarce resource controlled by the federal government. See Chapter 3.
13 http://www.fcc.gov/guides/obscenity-indecency-profanity-faq
14 For a comprehensive explanation of the FCC's authority, rules, regulations, and interpretations regarding broadcasters, see *The Public and Broadcasting*: http://transition.fcc.gov/mb/audio/decdoc/public_and_broadcasting.html#_Toc202587533
15 See Chapter 3, where common carrier status of wireline telephone systems is discussed.
16 For the full text of the 1996 Act, see: http://transition.fcc.gov/telecom.html
17 An interesting take by Consumers Union on why the 1996 Act failed to create as much competition as it promised is at: http://www.consumersunion.org/telecom/lessondc201.htm
18 Portability means you can keep your phone number when you change carriers or switch among cellular, VoIP, and landline systems.
19 The FCC has jurisdiction over common carriers, but not information services, which is how the Internet is classified. See Chapter 3, section "The regulation dilemma".
20 http://www.ushistory.org/declaration/document/
21 With apologies to Lewis Carroll, *Alice's Adventures in Wonderland*.
22 The uprisings are looked at in Chapter 6 in the context of social media and mobile communications.
23 A firewall is a rule-based device that can allow or deny access to a source. A proxy server is an intermediary between a requester and the server that actually can fill the request; the latter is hidden from direct access. The proxy also can deny access to the relevant server.
24 As quoted by David Barboza in "New China Search Engine Will Be State-Controlled," *New York Times*, August 14, 2010.
25 https://www.torproject.org/index.html.en
26 For a more detailed description, see James Glanz and John Markoff "U.S. Underwrites Internet Detour Around Censors," *New York Times*, June 12, 2011: http://www.nytimes.com/2011/06/12/world/12internet.html

27   Skip Priest, Congressman, Washington State: http://thinkexist.com/quotation/the-effort-here-is-for-once-and-for-all-to-give/1523137.html

28   Judge Thomas M. Cooley (1824–1898), as quoted by future Supreme Court justices Samuel D. Warren (1852–1910) and Louis D. Brandeis (1856–1941) in "The Right to Privacy," *Harvard Law Review*, IV, no. 5 (Dec. 15, 1890): http://groups.csail.mit.edu/mac/classes/6.805/articles/privacy/Privacy_brand_warr2.html

29   Private property is mentioned in Amendment V—"nor shall private property be taken for public use, without just compensation." Read more: Amendments to the Constitution of the United States: Infoplease.com http://www.infoplease.com/ipa/A0749825.html#ixzz1QOu5GFn6

30   FOIA: http://www.foia.gov/about.html

31   PA: http://www.justice.gov/opcl/privstat.htm

32   RFPA: http://epic.org/privacy/rfpa/; GHB http://business.ftc.gov/privacy-and-security/gramm-leach-bliley-act

33   ECPA: http://legal.web.aol.com/resources/legislation/ecpa.html

34   HIPAA: http://aspe.hhs.gov/admnsimp/pl104191.htm

35   The text of Section 230 is at: http://www.citmedialaw.org/section-230

36   The complete FIPP is at: http://www.ftc.gov/reports/privacy3/fairinfo.shtm

37   A Radio Frequency Identification Device (RFID), such as EZ-Pass, automatically charges your account for tolls and records the time you passed through a particular toll point. Their use has expanded to include automatic payment at some parking facilities and other venues.

38   Latanya Sweeney: http://dataprivacylab.org/people/sweeney/

39   Alessandro Acquisti: http://www.heinz.cmu.edu/~acquisti/

40   Ralph Gross: http://www.ralphgross.com/

41   Source of the three examples: http://epic.org/privacy/reidentification/#law

42   One highly successful media service that uses this technique is DoubleClick, now owned by Google, that unparalleled amasser of data (http://www.google.com/doubleclick/).

43   The information a site collects becomes the property of the site. Deleting a cookie does not delete it.

44   For FCA full text as amended see: http://www.arentfox.com/publications/index.cfm?content_id=879&fa=legalUpdateDisp

45   For FERA full text see: http://frwebgate.access.gpo.gov/cgi-bin/getdoc.cgi?dbname=111_cong_bills&docid=f:s386enr.txt.pdf

46   For WSRCPA full text see: http://www.sec.gov/about/laws/wallstreetreform-cpa.pdf

47   For OSH Act full text see: http://63.234.227.130/dep/oia/whistleblower/index.html. For information about OSHA, visit: http://www.osha.gov/

48   Bill Keller, "Secrecy in Shreds," *New York Times Magazine*, April 3, 2011. For the full text of the Papers and some historical background, visit: http://www.archives.gov/research/pentagon-papers/

49   http://WikiLeaks.org/

50   As reported by Jennifer Epstein in *Politico*, November 29, 2010 http://www.politico.com/news/stories/1110/45667.html. Also, view King's interview on the *Today Show* at http://www.youtube.com/watch?v=gkELILA47bo

51   From CBS News *Early Show* interview: http://www.cbsnews.com/video/watch/?id=7099181n

52   As reported by Stephen Lendman, in the *Baltimore Chronicle*, January 9, 2011: http://baltimorechronicle.com/2011/010911Lendman.html

53   As quoted by Vikas Bajaj in "India Puts Tight Leash on Internet Free Speech," *New York Times*, April 27, 2011: http://www.nytimes.com/2011/04/28/technology/28internet.html

54   From *The Gazette of India: Extraordinary*, Part II-Sec. 3(i), Notification, New Delhi, April 11, 2011: http://www.mit.gov.in/sites/upload_files/dit/files/GSR314E_10511(1).pdf

55   For the full text of CPBRA see: http://kerry.senate.gov/imo/media/doc/Commercial%20Privacy%20Bill%20of%20Rights%20Text.pdf

56   For more detail, see: http://www.democraticmedia.org/consumer-groups-welcome-bipartisan-privacy-effort-warn-kerry-mccain-bill-insufficient-protect-consum

57   For the full text of LPPA see: http://franken.senate.gov/files/docs/110614_The_Location_Privacy_Protection_Act_of_2011_One_pager.pdf

58   For the full text of the complaint, see: http://epic.org/privacy/facebook/EPIC_FB_FR_FTC_Complaint_06_10_11.pdf

59   For the full text of the Act see: http://leahy.senate.gov/imo/media/doc/BillText-ElectronicCommunicationsPrivacyActAmendmentsAct.pdf

5

What's mine is whose?—intellectual property

Intellectual property is an important legal and cultural issue. Society as a whole has complex issues to face here.[1]

Americans have been selling this view around the world: that progress comes from perfect protection of intellectual property. Notwithstanding the fact that the most innovative and progressive space we've seen—the Internet—has been the place where intellectual property has been least respected. You know, facts don't get in the way of this ideology. This is what we've been selling.[2]

Intellectual property—what and why

Pursuits that produce such works as inventions, art, and literature are products of the mind. These terms are broadly defined, with no judgment as to the quality or merit of the works implied. The question is, should the creators of these works be given some form of control over them? The answer in most countries has been yes, primarily on the grounds that doing so is an incentive to potential creators and so encourages creativity.

The mechanisms for providing control are based on property rights. When property can be owned, the owner can be given legal protection and control over it. That applies whether the property is real (physical) or intellectual.

Thoughts alone are not intellectual property. As long as a creative work is solely in the mind, it's ephemeral and from a property perspective, doesn't exist. The moment it's expressed in a tangible form—writing, drawing, painting, plan, graphic, recording, representation, device, machine, or instrument—it takes on the dimension of property.

> **Property and the Constitution**
>
> The Constitution specifically notes property only in the Fifth Amendment, which says in part "nor shall any person . . . be deprived of life, liberty, or property, without due process of law; nor shall private property be taken for public use, without just compensation." Though intellectual property can be assumed to be covered, the focus here is on the taking of physical property and not on general property rights.

Information resulting from a creative process qualifies as property. When a schematic is drawn describing a product or process, when a computer program is coded, when a news report is written, when a blog is posted, the information becomes intellectual property.

There are three kinds of legal protection for intellectual property—copyright, patent, and trademark. Each is aimed at different sorts of intellectual property and each has its own set of rights and rules. In the US, authority to establish conditions and protections for such property is given to the Congress by the Constitution:

> The Congress shall have Power . . . To promote the Progress of Science and useful Arts, by securing for limited Times to Authors and Inventors the exclusive Right to their respective Writings and Discoveries.[3]

Copyright—a little history

Following the American Revolutionary War (1775–1783), several states enacted some forms of copyright, but uniform protection in the US didn't begin until passage of the federal *Copyright Act of 1790*. The first sentence defined its purpose:

> An Act for the encouragement of learning, by securing the copies of maps, Charts, And books, to the authors and proprietors of such copies.

To secure a copyright, a copy of the work had to be deposited "in the clerk's office of the district court where the author . . . shall reside." The clerk would record the deposit and issue a seal of copyright notice, protecting the work for an initial term of 14 years renewable for another 14 years.

Over the years, copyright law was revised several times, primarily extending term length and adding to the list of what was copyrightable. In 1831 the so called *First General Revision of U.S. Copyright Law* lengthened the initial term to 26 years and added music in printed form to the coverage.

In 1909 both the initial term and the renewal period were extended to 28 years, but the works had to be published and have a notice of copyright affixed to them. Art, prints, reproductions, and photographs were added to the list of copyrightable items.

The next major revision was the *Copyright Act of 1976*. It prolonged the term of the copyright to life of the author plus 50 years, or 70 years for corporate holders. It removed the necessity of publishing the work, requiring only that it be in fixed tangible form. It also was more specific about the types of work that were copyrightable and the exclusive rights granted to copyright holders. Covered were works of literature, music and lyrics, drama, dance, graphics, photography, sculpture, movies, audio, and video. Holders were given the right to sell and lease, copy, perform, and display their works.

Most significantly, non-infringing use of copyrighted material was provided for by what is called *fair use*, which grew out of several court decisions. Specifically, copyrighted material can be used in limited amounts if it: is involved in a parody of the original work; does not deprive the original creators of income for their work; contributes to the creation of new work, such as a review, a critique, a news report; is used for educational purposes, including for classes, research, or other scholarship. However, anything copied has to be attributed and marked as quoted.

Flexibility was added by noting factors to consider in deciding whether fair use applied: the nature of use—commercial, nonprofit, or educational; the nature of the work—fictional or factual[4]; the amount copied compared to the size of the entire work and the significance of the copied portion regardless of size; the effect of use on the value of the original work or its market potential value—this one often carries the most weight.

The Act also removed the requirement that a work had to be registered to be copyrighted. As soon as material is in tangible form it's automatically copyrighted. Copyright exists solely by dint of creation and expression; you don't even have to add ©.

That might not be enough, though. If an ownership dispute comes up, you need to prove you were the first to express the idea. For that you can rely on testimony of witnesses, friends, and associates and you can mail a witnessed, dated, notarized copy of the content to yourself via registered mail and keep it unopened, but neither of those is reliable. You may have to prove your case in court, where testimony might not hold sway and can be countered by opposing testimony, and because mail is not tamperproof.

The safest way to prevail in a copyright ownership dispute is to register the content with the Copyright Office.[5] For a small fee, that is definitive proof that you were first to copyright, hence are the legitimate holder should someone accuse you of infringement or if you need to prove infringement by someone else.

Copyright life was dramatically extended by the *1998 Copyright Term Extension Act*. It increased the length of copyrights to 70 years past life of the creator, and for corporate ownership to 120 years after creation or 95 years after publication, whichever comes first. Arguments against the tenor of copyright laws often cite those terms as stifling creativity and innovation rather than incentivizing them.[6]

In fair use, vagueness abounds

The line between fair use and infringement can be fuzzy. Copyright holders have exclusive rights to derivative works, yet fair use may allow others to create such derivatives as reviews, critiques, and reports. But if too much of a movie is shown in a review, if too many passages of a story are quoted in a report, if a critique includes photos of participants, infringement can be claimed.

How much is too much? That's the question.

A contradiction that's not

A creative work can be copyrighted only if it's in tangible form. Yet it's the intangible that's protected, not the tangible. Though seemingly contradictory, it's not, as some examples show:

- It's not the paper and ink of a book that's protected, it's the story expressed.
- It's not the film that's protected, it's the plot, acting, and presentation.
- It's not the CD that's protected, it's the music and performances on it.

The IT effect

> *Copyright law has for a very long time been a tiny little part of American jurisprudence, far removed from traditional First Amendment jurisprudence, and that made sense before the Internet. Now there is an unavoidable link between First Amendment interests and the scope of copyright law. The legal system is recognizing for the first time the extraordinary expanse of copyright regulation and its regulation of ordinary free-speech activities.*[7]

There are two basic kinds of copyright violation—claiming too much of another's work as your own and wholesale copying, whether for personal use or distribution. Early on, wholesale copying was quite difficult. Making a copy was a tedious process, whether by hand or by taking photos. After copying machines arrived it got a little easier, but copying many pages still took a long time.

Importantly, machine copies were not as good as the originals and so were easily identifiable as duplications. That provided a kind of security in that they couldn't be passed off as originals, but of course, the material still could be sold or the information usurped and incorporated into another work.

Pre-Internet, most copyrights were owned by businesses and professionals. Jumping ahead to the present, we see that anyone who "publishes" anything via the Internet or other digital communications methodologies automatically becomes a copyright holder. But the same digital processing technology that makes such publishing cheap and easy to do also makes it simple to produce perfect copies that are indistinguishable from the originals—so there goes copyright protection.

Consumer digital IT, which started slowly but rapidly expanded in scope and capability as costs plummeted, is particularly illustrative of how convenience advanced and protection declined. Let's take a brief look back to see what happened.

- The first consumer *CD player* in US, the Sony CD 101, debuted in 1982. It sold for $900, which is over $2,030 in 2010 dollars.
- The first commercial *CD recorder* hit the market in 1991. The Denon DN770R was priced at $20,000. A blank recordable disc cost $40. In 2010 dollars, those amounts are about $32,000 and $64, prices that put them out of the reach of most consumers.
- Phillips followed six years later with a consumer version, the CDR 870. Retail price was down to $774, over $1,000 in 2010 dollars.
- Sony marketed the first *DVD player* in the US in 1997. The DVP-S7000 sold for $970, over $1,200 at 2010 rates.
- Three years later, Panasonic produced the first consumer *DVD recorder*, the DMR-E10, with a list price of $4,000 and $35 for blank discs. Those prices translate to about $4,250 and $45 in 2010 dollars.

As technology progressed, CD and DVD players and burners became standard peripherals for desktop and laptop computers. At the same time, hard disk capacity surged and physical size and price plummeted.

The first hard disk for personal computers was marketed in 1980 by Shugart Technologies. The ST506 was about 5" high, 6" wide, 8" deep, had a capacity of 5 megabytes (MB) and sold for $1,500 ($4,000). Now you can buy a Seagate Technologies 1" 8 gigabyte (GB) hard drive for less than $115.[8]

USB consumer flash drives arrived in 2000 and followed a similar capacity/cost trajectory. The first such device, an 8MB *thumb drive,* was produced by Trek Technology. One 2011 model Kingston Data Traveler flash drive, physically about the same size, has a capacity of 256GB.

In parallel to these hardware developments, data compression algorithms improved substantially, enabling significant reduction in file size. Combined with growth in broadband service, Internet transfers of even huge files as well as real-time audio and video streaming became exceedingly feasible. For very little cost and in very little time, all sorts of information can now be uploaded, widely distributed, downloaded, and copied by anyone with basic equipment and an Internet connection.

There is every reason to believe these trends will continue. So once again, improvements in IT have led the way— this time to much greater convenience and much greater opportunity for copyright infringement.

> ## CDs, DVDs, the Internet, and the year 1841
>
> Digital media and digital transmission systems represent information by bit sequences, so digital data are discrete and finite. Audio, video, and music in their native states are analog, hence continuous and infinite. An infinite amount of information can't be directly converted to a finite form. Instead, the analog source must be sampled to digitize it.
>
> In 1841, French mathematician Augustin-Louis Cauchy (1789–1857) developed a sampling theorem to convert analog streams into numbers. It was the basis for subsequent work by Harry Nyquist (1889–1976) and Claude Shannon (1916–2001) that led to current sampling and compression algorithms that make digitization of analog data for recording and transmission over the Internet practical.

Legislators try to cope

Definitions in the laws, even those that seem to be general, become insufficient and even inapplicable as IT changes, and the speed of change is too fast for the laws to keep pace. In much the same way as closed regimes play a cat and mouse game with those who try to breach access barriers, a similar game is played between those who try to strengthen the means of copyright protection and those who try to get around restrictions. The speed of IT development and its features that clever users can take advantage of have created an escalating tug of war.

The comprehensive US *Copyright Act 1976* predated not only today's vast use of PCs, smartphones, the Internet, and the Web, but also the development of such commercially successful information carriers as the audio cassette tape recorder, video tape recorder, CD and DVD burners, MP3 compression, and content streaming.

Tape recorders made volume copying and distributing of analog source material, copyrighted or not, cheap and easy. CD and DVD burners, compression, and streaming did the same thing for digital material. Despite attempts to rein in illegal copying and downloading, people have shown that if it's easy and cheap to do, they will do it. That made these IT developments the source of increasingly heated disagreements about the balance between copyright protection and creativity suppression, what fair use is and should be, and how to treat uses that technically fall under the legal definition of copyright violation though logically they should not.

Once consumer tape recording took off, content providers found their business models threatened. In response, they pursued legal actions against companies that produced the machines and pressured Congress to enact protective measures. The result was the *Piracy and Counterfeiting Amendments Act of 1982* (PCA Act)[9] aimed at stopping "Trafficking in counterfeit labels for phonorecords, and copies of motion pictures or other audiovisual works" by imposing fines of up to $250,000, prison terms of up to five years, or both.

Disputes about the fairness of this Act revolved around overreaction to what should be fair use copying and overzealous protection of content. One watershed case in that regard was *Sony Corp. of America v. Universal City Studios, Inc.,*[10] in which the Studios cited Sony's Betamax video recorder as providing the means for copyright infringement. The case, brought before the Supreme Court, was decided in Sony's favor.

The Court's two major conclusions were that recording TV shows for time shifting was fair use and that manufacturers of such recording equipment could not be held liable for violations just because the equipment could be used to infringe on copyrighted material. That result turned out to be extremely beneficial for both sides. Equipment manufacturers saw a boom in sales and movie companies made heaps of money from the sale of movie videos. Still, under the terms of the law, individuals could be held liable for copying a movie they legitimately bought, even if for their own use.

A decade after the PCA Act, it was refined to define copyright violation as high volume copying—making at least ten copies with retail value of more than $2,500 within six months. That quieted the controversy somewhat, but not for very long. In 1993, IT leaped ahead of the laws when David LaMacchia, a student at MIT, created an online bulletin board that allowed participants to copy copyrighted software and video games from each other's machines. Once again, creators, producers, and distributers found that a disturbing threat to their business models and revenue streams.

The bulletin board proved so popular that within two months millions of dollars' worth of copies were made.[11] Was LaMacchia guilty? Existing anti-copying laws focused on infringements that resulted in a financial gain for the infringer. That was his twist—he didn't charge for the software, didn't receive any money from copying activity, and didn't keep a repository of the software, all of which resided on the machines of various members. That meant that he was innocent under the law because "In addition to demonstrating willfulness, the government must prove that the defendant engaged in an act of infringement 'for purposes of commercial advantage or private financial gain.'"[12]

Not being able to successfully pursue LaMacchia in the courts, prosecutors tried a different tack, charging him with committing wire fraud. That 1994 case[13] failed too, because of the precedent of a1985 Supreme Court decision[14] that concluded that theft or fraud didn't apply to intangible intellectual property—bit transmission of files over a network did not render them tangible.

By 1997, Congress saw that the growing capabilities of digital network communications provided a ready means for employing schemes that could easily skirt copyrights simply by detouring around "commercial advantage or private financial gain." So they removed that requirement with passage of the *No Electronic Theft Act* (NET Act).[15]

Two other issues remained: the variety of rules enacted by other countries; and content transmissions over digital networks being outside the scope of copyright law.

WIPO to the rescue   Copyright laws are national. No nation has to obey the laws of another. As it happens, many do have reciprocal agreements and others do not. Afghanistan and Laos have no copyright laws at all. To be on the safe side, creators of intellectual property have applied for copyrights in many countries, a rather tedious process.

Reacting to that dilemma and issues of dealing with copyright protection in the digital IT world, the World Intellectual Property Organization (WIPO), crafted the *WIPO Copyright Treaty* to focus on international copyrights. It was agreed to by member countries of WIPO in 1996.[16]

The Treaty simplified obtaining international copyrights by creating a single application that covered all the participating countries. It also contained a number of protections for IT-based copyrighted content regardless of how it was distributed and, importantly, banned the use or provision of methods to disable copy protection. For balance, it had a safe harbor stipulation that shielded content distributors from copyright violation or abetting violation if they removed infringing content when asked to do so by copyright holders.

Some parts of the Treaty were decried as unwise or unfair. In particular: banning the disabling of content protection in all cases, even for fair use, was called too restrictive; a common policy for all signatories, whose IT capabilities and economic development were in very different stages of sophistication, was felt to be unworkable; and lengthy terms of protection were opposed as stifling creativity—almost all signatories have some form of life plus 50 to 70 years. Of course, these objections were disputed by WIPO and supporters of the Treaty.

The DMCA   The WIPO Treaty notwithstanding, copyrights remain national prerogatives. Until a country crafts its own laws incorporating Treaty provisions, they don't take effect. Several countries did that fairly quickly. Others delayed much longer and a few others have not moved at all.

In the US, implementation waited until passage of the simultaneously loved and reviled *Digital Millennium Copyright Act of 1998* (DMCA). Among its many clauses, it incorporated two major provisions of the 1996 WIPO Treaty: one made illegal the creation, distribution, or use of tools to break media copy protection, punishable by fines of up to $500,000 and five years in prison for the first offense; the other granted safe harbor for Web site operators.[17]

The DMCA had wide-reaching effect. One result with many reverberations was that it spelled doom for Napster, the groundbreaking service that made MP3 music file sharing and swapping simple and easy.[18] Formed in 1999, Napster used a central server that held a directory of members' computers and their MP3 files, but not the files themselves. A member could search the server to locate a file and download it from the computer that held it—a so-called peer-to-peer service.

Napster's popularity and membership grew rapidly, with huge numbers of songs being shared without permission or copyright holder recompense. That growth brought it to the attention of individual artists, recording companies, and the Recording Industry Association of America (RIAA), which represents record labels and distributors.

Lawsuits followed. That Napster didn't directly hold or copy any file was the hook that would have protected it from copyright infringement. Then along came the DMCA abetting clause. The heart of the suits was that Napster was abetting copyright infringement by not removing access to infringing material, a DMCA and Treaty requirement for safe harbor. That was something Napster couldn't do because of the interwoven nature of the service's members and the fact that the same files were likely to be on many different computers over which Napster had no control. Napster lost the suits and shut down in mid 2001. It has since re-emerged in significantly different form as a pay-for-content online music store, currently owned by Best Buy.

DRM invigorated   *Digital Rights Management* (DRM) refers to a set of methods for controlling what can and cannot be done with digital content. The DMCA, by its prohibition of using or distributing tools to break copy protection methods, gave considerable boost to DRM implementations, since disabling DRM code fell under that prohibition.

Beloved by copyright holders, they claim it to be not just a necessary tool to prevent infringement, but one that has the added bonus of protecting against viruses. It is vigorously opposed as interfering with legitimate and logical personal rights of content purchasers, and its ability to defend against viruses is disputed.[19] Let's see how the claims stack up.

DRM has two main features: content encryption that requires user authentication before it can be decoded; and special digital marks that implement copy management, preventing systems from making unauthorized copies. These seem entirely reasonable, but further investigation turns up some serious issues, the two biggest being control and privacy.

DRM can prevent copying a disc you bought to your own portable player and can specify which equipment the disc will play on at all. DRM can dictate which reader must be used for the e-book you purchased. DRM can specify the app provider you must use with your smartphone. DRM can force you to go through the trailers on a video disc and prevent you from skipping warnings before reaching the menu. In other words, you are not free to use legitimately purchased media and content as you like, even if those uses are within the law or fair use.

DRM authentication can require providing personally identifiable information, proscribing anonymous use. The information can be used to create consumer profiles and track usage patterns and preferences. Opponents see these as invasions of privacy—after all, how you use what you legitimately buy is nobody's business but your own, isn't it?

BitTorrent joins the fray   BitTorrent is a different take on peer-to-peer digital file sharing. As with the Napster model, files are located on many member computers, but unlike Napster there is no central repository or directory. Instead, members create small files called torrents that have brief descriptions of the files on their machines.

Members can find the files they want to download by searching for torrents using any conventional search engine.

Another key difference from Napster is that files are not handled as wholes. When a file is to be downloaded, it's broken into small pieces. Since many members will have the same file, different pieces will come from different members, not from just one. That eases the burden that a single source would otherwise have, especially when many members want the same file. It also means that download speed doesn't decrease noticeably when multiple users download a file at the same time. In addition, transmitting small pieces rather than an entire file requires much less network bandwidth, a feature that makes BitTorrent popular for downloading large files like movies.

As downloads take place, BitTorrent sources spread. Each newly downloaded file becomes a source for other members, creating a seeding effect that multiplies the sources of a file as others download it too. The more popular a file the more widespread it becomes, further spreading the sharing burden.

The BitTorrent company[20] develops and maintains BitTorrent client software and makes it available without charge or usage fees. Because it's the software that enables all this file sharing, BitTorrent has been accused of abetting copyright infringement, but the software just assists downloading files, which are not necessarily copyrighted. That's a key point deserving of emphasis.

BitTorrent client software is a tool that can be used to efficiently download files, period. Whether those files are legally or illegally downloaded is up to the users, not the BitTorrent company. In the same way that a crowbar manufacturer can't be held responsible if a person uses it to break into a house, a weed killer manufacturer can't be sued if someone uses it to poison a rival, and Microsoft isn't liable if a person uses Word to write a blackmail letter, so a file transfer software creator isn't legally responsible if its product is used illegally.

After all, file downloading in itself isn't illegal. Furthermore, the company stores no content, not even the torrents, has no control over or claim on anything downloaded, and does not track users. As a result, BitTorrent has so far eluded prosecution even though a significant number of illegal file downloaders do use the software.

It remains to be seen if the laws will be revised to close what many content creators and copyright holders consider to be a loophole and others believe is just a poor interpretation of what abetting means. As you can imagine, crafting such a law is a tricky business at best, and likely to suffer at least as much from rights violations as it would succeed in preventing abuse.

What do you think of this? *Sample at your peril*

Jazz has a long and varied history of musicians borrowing from, improvising off of, and creating derivative works based on the compositions and performances of each other. That has resulted in a rich, impressive body of work for which there was never a question of copyright infringement.

Music sampling, wherein snippets of sound recordings are processed and combined in a variety of ways—for rap composition background, incidental to scenes in movies, and the basis for newly derived works—has not been treated as kindly. Infringement for the slightest use is considered proof of copyright violation, pursuable by the music's copyright holders—writers, publishers, recording companies, and artists. Payments demanded for permissions, even for trivial use, can be in the thousands.

Fair use, which typically allows permission-free use of copyrighted material if the amount taken is small in relation to the whole, is substantially transformed, and doesn't cause significant financial harm to the owner, seems to be ignored when it comes to music sampling, even when the snippets taken are just a couple of seconds.

Unlocking content

Despite the long life of a copyright, it isn't forever. Once it expires, the material enters the public domain, where it is free to everyone without restrictions. Life plus 50 to 70 years is a long time to wait, though. There is much sentiment to the effect that copyright law stifles creativity because the term is so long. Worse, use can be blocked even when the owner is dead and has no heirs to inherit the copyright, or when the material is out of print, non-circulating, and the holder can't be found.

On top of that, we've become accustomed to easy access, copy, paste, post, and the like delivered by the current capabilities of IT, especially the Web, the Internet, and huge numbers of applications. That has led to a boom in creativity as people build on each other's works. The issue is that to be legal, these activities require explicit permissions when copyrighted material is involved. Disputes arise when fair use is questioned, and they can be quite vehemently argued.

Movements in opposition to current copyright law and practice have slowly begun to grow. Two significant efforts to provide alternatives have taken related paths: the *GNU General Public License* (GGPL), also called a *copyleft license*, and the *creative commons license* (CCL).

In the words of the GNU Organization, the GGPL "is a free, copyleft license for software and other kinds of works."[21] If you make your work available under a GGPL, anyone can use it, modify it, copy it, and so on, but if you make your derived work available, you also must do so by a GGPL. This has been a great help to open source and free software developers who want to formalize simple usage agreements for their work.

The CCL, inspired in part by the GGPL, is similar but gives creators more control. In their words: "Creative Commons is a nonprofit organization that develops, supports, and stewards legal and technical infrastructure that maximizes digital creativity, sharing, and innovation." Their tools "give everyone . . . a simple, standardized way to keep their copyright while allowing certain uses of their work."[22]

The CCL works within existing copyright laws to give you flexibility in specifying how your work can be legally handled, which they call a "some rights reserved" copyright, rather than the *all rights reserved* typical of standard copyrights. With it you can choose which permissions to grant for various uses of your material— copying, distributing, editing, remixing, adding to, and creating derivative works— and you do that in one stroke for all potential users rather than for each specific instance.

You still retain your copyright and get credit for content that's used by others. You can choose to allow or disallow commercial use. To keep the process going, if you allow derivative works, those creators are required to issue their works with a CCL under the same terms as the material they were derived from.

Copyright holders fight infringement   Corporate copyright holders, especially those in the music, movie, and software industries, have adopted the word *piracy* to describe illegal downloading, copying, and distribution. For centuries, piracy has meant acts of violence, attack, and hijacking by threatening parties out to reap bounty. Seized on by the media, corporations, politicians, and countless other individuals, the term casts an ominous aura over infringement.

Such overblown usage demonizes violators, often not the best way to approach a problem. After all, we've seen usage that's technically illegal but can be viewed as being the impact of improper or poorly interpreted and implemented laws. Not every infringer is evil, out to attack, or reap a bounty—their activities range from benign to egregious. A more appropriate term for illegal copying is theft—simple and straight to the point.

Whatever it's called, there's no doubt that illegal copying has been going on for some time, jumping considerably in the digital age. Industry, having tried the relatively straightforward infringement-thwarting methods of encryption, expiration dates, dongles,[23] and education, has seen them work well for a time and then lose effectiveness. That happened as hackers and developers of work-around sites took advantage of the Internet to facilitate illegal downloading, largely nullifying those schemes.

In reaction, the industry ratcheted up the force of its moves by instituting lawsuits seeking enormous monetary judgments to punish violators and warn off potential infringers—the RIAA alone has sued over 20,000 people. That tactic also worked for a while, but after an initial shocked reaction and some modest success, a growing backlash of negative publicity reduced that practice considerably.

What do you think of this? *Justified or overkill?*

The first major lawsuit for file sharing copyright infringement was brought by several record labels against Jaime Thomas. She was found guilty in 2007 for sharing 24 songs and was fined $220,000. Granted a retrial in 2009, the decision again went against her and the penalty was increased to $1,920,000. Later it was reduced to $54,000, but the plaintiffs refused to accept that reduction, so in 2010 a third trial was held. The result, a $1,500,000 judgment against her. That's $62,500 a song.

The digital difference. "The basic principle in the law is that you have to distribute actual physical copies to be guilty of violating copyright. But recently, the industry has been going around saying that even a personal copy on your computer is a violation" (Ray Beckerman, a New York lawyer, quoted in: http://www.washingtonpost.com/wp-dyn/content/article/2007/12/28/AR2007122800693.html).

The illegality of infringement is not in dispute. The issue is, regardless of how it's carried out, at what point does monetary punishment cross the line between appropriate compensation (and discouragement of the practice) and exorbitant onerous judgments? The question: does the penalty fit the crime? Is discouraging others sufficient reason for draconian punishment of an individual?

Lately, a new approach has been devised that gets the message across without requiring harsh judgments or complicated technology. Named the *Copyright Alert System* (2011), it calls for ISPs notified by content owners of illegal downloads and file sharing to send warning emails to infringers and slow down their connections for a time. Subsequent violations would result in more warnings, further slowdowns, site blocks, and eventually ISP account shutoff. The hope is that when service gets slow enough, downloading would take too long to be worthwhile, or at least be so annoying as to discourage the practice.[24]

Recent attempts by Congress to address illegal downloading are currently in limbo. One bill proposed by the Senate, called the *Preventing Real Online Threats to Economic Creativity and Theft of Intellectual Property Act* (Protect IP Act, or PIPA) is a revised version of an earlier proposal called the *Combating Online Infringement and Counterfeits Act* (COICA). The latter gave the US Attorney General the power to petition the DNS system to block access to infringing sites by freezing their domain names, to have ad networks stop working with infringing sites, and to require credit card companies and other payment processors to refuse to handle payments for those sites.

Because COICA took a broad strokes approach that could inadvertently harm innocent sites and operators, the bill was revisited with the intent of aiming more directly at stopping sites that continue to violate copyrights. Of course, since almost all those so-called rogue sites are based in countries out of the reach of US laws, that is easier said than done.

PIPA did narrow COICA's definition of a rogue site, but it added two more provisions: search service providers would be required to not show results for any sites whose domain names were frozen; and copyright holders could directly apply for court orders to invoke the ad network and payment processor provisions without having to wait for government action. So in fact, PIPA is more severely restrictive than its predecessor, though not as draconian as the bill formulated by the House of Representatives, called the *Stop Online Piracy Act* (SOPA).

Beyond provisions similar to those of PIPA, SOPA defined copyright violations so broadly that even brief snippets or usages that comply with the fair use terms of the DMCA and WIPO could be considered infringement. That could easily affect sites that supply reportage, reviews, and critiques. More than petitioning the DNS to block infringing sites, under SOPA, ISPs can be held responsible for barring access.

These are substantial increases in severity, counter to the DMCA safe harbor provisions that hold sites harmless if they remove infringing material after being notified, and ISPs that are not deemed responsible for supplying access to such sites. In other words, the distinction between a site, an access provider, and persons who post possibly infringing material gets swallowed up.

Both the Protect IP Act and SOPA are fiercely supported by the music and movies industries that traditionally oppose anything different from the status quo, and just as fiercely opposed by access providers, search services, and a host of others concerned with Congress' attempts to muzzle the Internet and start the US down the road to Web site censorship, a slippery slope at best. The supporters of these acts, heavy donors to legislator campaign committees, have the advantage of being much more influential in Congress than the opposition.

In late January 2012, Wikipedia closed its site for 24 hours, replacing a search request with a page stating its opposition to SOPA and PIPA. Other sites posted banners also opposing the laws. Tumblr was quite vociferous in its protests and Twitter was active with mostly anti-SOPA tweets. The growing outcry against the acts, not all of which are based on the facts of the situation, has caused some lawmakers to change their minds about supporting the laws. That's one more demonstration of the non-organized organizing power of the Internet.

As of this writing, PIPA, already passed by the Senate, was put on hold and is undergoing modifications. SOPA, not yet passed by the House, is also being reconsidered. Either or both may eventually return in different forms, be replaced by something else, or be dropped altogether. If both acts subsequently pass in some form, the next step is for the House and Senate to reconcile their two versions, after which the combined bill would be voted on and, if passed, move on for presidential signature or veto.

Defenders of the bills' concepts state that they are minimally invasive means of diminishing theft of intellectual property, that they are aimed solely at foreign rogue sites, do not interfere with legitimate activity, and do not inhibit any of the legitimate uses of the Internet that we have come to enjoy. Opponents say that the way the laws are crafted, there is a real danger that legitimate users will be ensnared, that unintended consequences will be harmful to legitimate sites, and that miscreants will find a way around restrictions anyway.

> When it comes to protecting copyrights, privacy, security, or anything else that requires some sort of restrictions, anything that makes it harder to violate protections also makes it harder for legitimate users. We never will be completely rid of infringements.
> A knotty question: Where should the line be drawn between protection and ease of use?

Regardless of which side you agree with, there is no doubt that infringement via the Internet is commonplace and that it does result in lost revenues for copyright holders. The issues are: Is there a level of infringement that is not overly onerous so that legitimate users do not have to be unduly restricted? Importantly, are there ways to encourage legitimate use and discourage illegal use that don't depend on laws at all?

Rethinking business models   Many of the methods used to protect copyrights and preserve traditional business models are so burdensome and restrictive that they have alienated legitimate users, pushing them to infringing alternatives to get around the illogical constraints of legitimate use. Then illegal downloading is only another click away. There's also the matter of content cost, which can be relatively high for legal purchasers and zero otherwise. Clearly, illegal infringement has grown even in the face of complicated protections and harsh punishments, so it's doubtful that increasingly punitive methods will work. More likely is a backlash.

The fact is that of the two approaches to achieving compliance—the carrot and the stick—almost all business model preserving methods are of the stick kind, clearly illustrated by PIPA, SOPA, and their predecessors. That is a significant part of the problem. Very few attempts have been made to take the carrot approach by developing business models that are complementary to the capabilities of IT, especially the Internet and the Web, in ways that damp down the urge to download illegally—easier access, more individualized choice, greater variety, and lower costs. In the end, that would benefit everyone, just as did the Supreme Court decision that rejected the suit against Sony for producing a consumer video tape recorder.

One example of a carrot approach is that of the band Radiohead.[25]

> In 2007 Radiohead made their new album, *In Rainbows*, available as a digital download for which you could pay whatever you thought appropriate, or nothing at all.[26] The album was self-released, circumventing the need for a record label intermediary that was typical of the business, and usage was not restricted. The album also was released as a CD, which was not free but had features not in the download.
>
> As it turned out, Radiohead received at least as much income from the pay-what-you-feel-like downloads as they expected from a traditional studio distribution. Since then, they have released other digital albums for download at much reduced prices compared to typical CD releases, though not for free. Neither approach was a happy event for the label studios, but worked well for the artists.

Other business models are slowly beginning to pop up. Here are some examples:

- In 2001, *Apple, Inc.*[27] began selling individual music tracks by download through their proprietary media player, iTunes. You can create your own mixes and "albums" by buying individual tracks from the iTunes store rather than having to take whatever a studio decided to assemble in an album. This highly popular system charged 99 cents a track, so the cost for an album's worth of music was in the ballpark of many CDs', but the mix was yours and you could download as few or as many tracks as you wanted. Now all sorts of content can be bought for various prices and some items are free.
- *Jamendo*[28] developed an innovative business model to distribute music published under a CCL. Venues such as restaurants, clubs, salons, spas, hotels, fitness clubs, department stores, boutiques—just about any business that uses music as background or as a service—pay subscription fees for audio streams that can be tailored to their needs. The overall cost to the businesses is relatively low because there are none of the performance fees that copyrighted music distributers charge. The site and the artists who supply music share the income.

Home listeners can stream, download, and share music for free under the terms of the CCL, provided they aren't using the music to make money.

- *Pandora*[29] is a music streaming site on which you have a choice of 40 hours of music a month for free in exchange for listening to short ads every so often, or unlimited hours of higher quality audio without ads by subscription for a relatively small annual fee. Listeners create their own "stations" by specifying a genre or artist, and the site streams music accordingly. The downside is that the site controls a station's playlist, though songs can be skipped and marked as liked or disliked, which tailors the selections somewhat. Pandora pays small royalty fees for the music that's streamed, so their operation complies with the law.

- *Rhapsody*[30] follows a similar model but is available only by subscription, which is fairly inexpensive if you listen to a lot of music but is about four times the cost of Pandora's fee service. However, Rhapsody's song selection is customizable and can be streamed to a wide variety of devices.

- *Magnatune*[31] is a service that combines the ideas of Jamendo, Pandora, and Rhapsody. There is a free version with limits and ads, a subscription service for commercial establishments, and an income sharing system for artists and the site.

- *Amie Street* was an attempt to create a pay-by-popularity model. A song would be offered for free at first. Then the more it was downloaded the more it would cost. The site focused mostly on unknown or little known artists and was growing in popularity when it was bought by Amazon and changed to an MP3 streaming service where little is free but fees are generally low.[32] Amie Street's founders, in turn, created *Songza*,[33] a free streaming site offering a variety of genres via "stations" similar to Pandora's. Ads bring revenues to the site.

These and other attempts at new models show that the long-time standard industry studio/artist/contract model is neither the only way to sell music and other digital content nor a model that will flourish as is, since it is at odds with the way IT in general and the Internet and Web in particular are increasingly used. Time will tell which of these models and those to come will fail, survive, or thrive. One thing is clear—preservation of the past is futile; innovation and creative thinking will lead the way.

Patent—a little history

Congress passed the first US federal patent law, the *Patent Act of 1790*, the same year as the first copyright law, under the same authority granted by the Constitution. It gave property rights to patent holders for a 14-year term.[34]

Patents were granted by the Secretaries of State and War, and the Attorney General, for what were rather vaguely called "useful and important inventions." Applications had to be accompanied by detailed descriptions and models that were either working or could be made working by an appropriately skilled craftsman.

Just three years later, the *Patent Act of 1793* clarified what could be patented—"any new and useful art, machine, manufacture or composition of matter and any new and useful improvement on any art, machine, manufacture or composition of matter"[35] It wasn't until 1952 that the word "art" was replaced by "process." That small change was to become the basis for granting a slew of controversial business process and software patents, something that's become hotly debated.

Early on, the Department of State was the overseer of patent applications, farming out evaluations to a variety of professionals. By 1836, the volume of applications and court cases for infringements had grown too large to be handled piecemeal. So, with the *Patent Act of 1836* Congress created the US Patent Office, a dedicated operation separate but adjunct to the Department of State, to handle applications and court cases and to improve the patent evaluation and granting process. The Office remained under the jurisdiction of the Department of State until 1849, when it was moved to the Department of the Interior. In 1925 it was moved again, this time to the Department of Commerce where it still resides.

The 1836 Act called for distributing all new patent descriptions to libraries throughout the country, spreading information much more effectively. That made it easier for creative inventors to find out about and hence improve upon existing work, a precondition for the innovative boom of the nineteenth and twentieth centuries. It also helped to cut down on the number of patent lawsuits, many of which had arisen because the infringer didn't know that their device or more likely, some of its essential components, had already been patented. Additional protection was given to patent holders by adding the possibility of renewing a patent for seven years.[36]

In the years to follow, several disputes arose over whether a device deserved patenting if its only features were trivial extensions of existing patents. That was settled by an 1850 Supreme Court ruling that being "new and useful" was not enough for an invention to be patentable—it also needed to be an improvement that was not obvious to the average person.[37] Though a fairly fuzzy requirement—after all, who's an average person, and many things that seem obvious after the fact were not so obvious beforehand—it still was a step in the right direction.

The ensuing years saw several small changes stemming from court decisions and rulings by the Patent Office. Some examples: patents were allowed for hybrid, cultivar, and mutated plants (1954); the scope of the Office expanded, reflected in its new name, the Patent and Trademark Office (1975); genetically modified bacteria were declared patentable (1980); life of a patent was changed to 20 years with no renewals (1994).

WIPO steps in   Current interpretation of patent law owes much to the *WIPO Patent Cooperation Treaty*, established in 1970 and modified several times since, the last in 2001.[38] It defines patents as "an exclusive right granted for an invention, which is a product or a process that provides, in general, a new way of doing something, or offers a new technical solution to a problem." It also specifies three tests for patentability:

> [the work] must be of practical use; it must show an element of novelty, that is, some new characteristic which is not known in the body of existing knowledge in its technical field [and it must have] an inventive step which could not be deduced by a person with average knowledge of the technical field.

In the US and several other countries, a fourth test is sometimes added: implementability. That test can be invoked to deny a patent for something that is completely of the imagination—a time travel machine, a process for turning lead into gold—that has no realistic chance of being implemented in the foreseeable future. Of course, should an invention of that sort be made to work, that would be a different story.

As with the WIPO Copyright Treaty, each country has its own laws, national in scope. When countries' laws are congruent with WIPO provisions, they are congruent with each other. That's the intent of the WIPO Patent Cooperation Treaty and that's when international recognition exists. To eliminate the burden of having to apply for a patent in each country, WIPO has a single application that all signatories accept, similar to the single copyright application.

Most countries of the world, especially those with open societies, respect each other's patents. That is not the case in most closed societies, the same ones for which copyright protection is sketchy. Even in cooperating countries, there are significant differences. As the WIPO patent treaty puts it, "In many countries, scientific theories, mathematical methods, plant or animal varieties, discoveries of natural substances, commercial methods, or methods for medical treatment (as opposed to medical products) are generally not patentable."

US patent law runs counter to much of that. Patents are granted for software and business processes ("mathematical methods" and "commercial methods") and for biologicals, including genes ("discoveries of natural substances.") We return to this controversial practice later.

What do you think of this? *My genes belong to whom?*

For certain patents, 1980 was a key year because of a 5 to 4 Supreme Court decision (*47 U.S. 303, Diamond v. Chakrabarty, No. 79–136*) that made genetically engineered bacteria patentable because such bacteria didn't occur in nature. That opened a floodgate of patent applications for whole genes with known functions and DNA sequences of partial genes with unknown functions, naturally occurring or not.

Two major outcomes that work against such patents: patent holders could require fees for using the genes in diagnostic tests and therapeutic treatments or could monopolize those markets; partial gene sequence patents could force researchers completing the sequences to pay fees for the partial sequences.

Recently, US Federal Southern District Court of NY judge Robert W. Sweet invalidated seven patents for the BRCA1 and BRCA2 genes whose mutations are related to breast and ovarian cancer (*Assoc For Molecular Pathology v. United States Patent & Trademark Office 09 Civ. 4515, filed 3/29/10*). The patents gave owner Myriad Genetics a monopoly for testing the genes, for which they charged over $3,000 a test. Sweet said the patents were improperly granted because they concerned a law of nature.

So the question: Should patents for parts of nature be allowed, whether or not those parts have been modified? Gene patent supporters say ownership rights encourage innovation; opponents say they stifle it.

Types of patents

There are three basic patent categories. Protection for any is the same—exclusive ownership rights for 20 years. Applications for all types must pass the three or four patentability tests.

- *Utility patents* cover the way works function. The most common, they apply to mechanisms, electronics, pharmaceuticals, computer programs, and business processes. In addition to an application fee, utility patents have an annual maintenance fee.
- *Design* patents, the next most common, cover the appearance of an item—the look, shape, colors, style, ornamental and aesthetic features—without regard to functioning. As an example, a laptop computer may have many utility patents covering the way its particular components and systems function; the case may have a design patent for the way it looks. There is an application fee for design patents, but no maintenance fee.
- *Plant patents* are the least common. They cover new varieties of botanical plants produced by hybridization, genetic manipulation, or mutation. They also have an application fee but no maintenance fee.

A patent may be granted after an application is filed with and reviewed by the Patent and Trademark Office. Between the filing and awarding of a patent, the work can be marketed with the note "patent pending." That gives notice to others that an application has been filed—first to file is the rule for deciding who gets a patent if there are multiple applications for what is essentially the same device.[39]

It's important to note that a patent is not required to market something. With enticing but scientifically unsupported claims, many unpatented non-functional or poorly functioning products are sold to an unsuspecting public—miracle gadgets claimed to double gas mileage but actually have no effect or may decrease it; bracelets to guard against or cure diseases that have no scientific basis or statistically significant tests to back them up; age-reversing treatments that reverse nothing; weight reducing contraptions that only reduce the weight of your wallet; and the like.

Of course, not all non-patented devices are fraudulent. Those that are usually depend on IT to spread word of their supposed virtues quickly and widely, seeking an initial wave of purchasers whose money they reap before closing shop and moving on to the next worthless product. Legitimate devices that will have a short market life, whose producers don't want to wait to work through the patent process, may market their products unprotected.

Many agencies provide some protection against fraud—the federal Food and Drug Administration, the Consumer Product Safety Commission, the Department of Transportation, the Highway Safety Traffic Administration, and the Commerce Department among others, along with various state departments, the US Attorney General and state attorneys general. All play various roles in consumer protection, but none of it has to do with whether a questionable product is patented or not.

Protection but not secrecy   As we've seen, patents grant property rights, the authority to exclude others from making, copying, or incorporating the items in other products for a finite term, 20 years or so from application filing date in much of the world. To avoid damping creativity by hiding the inner workings of devices or processes, patents invoke a transparency strategy to reveal details—an application must contain a complete description, which is made public. That way, product and process development can be carried out by going beyond where the patented work leaves off, without fear of inadvertent infringement.

Of course, as with most everything else, that provision has a dark side—making it easier for infringers to duplicate the device. And so, progress and evasion continue.

The IT effect

Every piece of software written today is likely going to infringe on someone else's patent.[40]

There is an ongoing debate about whether software and business processes, most of which are software implementations, should be protected by patent or by copyright. Indeed, some contend they should not be protected at all. The arguments favoring either form of protection are essentially the same—ownership rights incentivize further work; so does economic gain from competitive advantage; and both of these stimulate creativity and motivate innovation.

The arguments against either form of protection also are essentially the same—the process takes too long, especially relative to the short useful life of software; the expenses of copyright or patent searches are high relative to the value of most software products; and patented software application details aren't published for 18 months after submission, by which time other developers may have already written infringing code unknowingly.

It seems there's not much to choose from in deciding on whether to patent or copyright software. In the end, it comes down to the principle conceptual difference between the two, which lies in what protection means for each: in broad terms, copyrights protect expression while (utility) patents protect function. But it's not always clear-cut.

If one computer program mimics the operations of another but differs in some internal design aspect, does that mean it's not patentable since the functioning is the same, but it is copyrightable because the expressions differ? And how much difference is necessary? Does fair use apply? Does coding have to be the same line for line, at least for substantial parts of a program, for copying to be inferred, hence copyright disallowed? If functionality is the same regardless of coding, is that enough to deny a patent?

However these issues are dissected, software patent protection is stronger than software copyright protection. For example, if the software implements a particular algorithm, patent protection would prevent others from encoding that algorithm even if the program that does it is written differently. That's not the case with copyright, again because it's not the idea that's protected, it's how it's expressed—different expression of the same functioning is allowed. That's much less protection for the creator of the algorithm.

Patent violation is another story. Illegal downloading, copying, and distribution methods commonly used to skirt copyrights are not relevant to most patents because their details are made well known and readily available, so they do not have to be purloined. The ability to detect violations varies, though.

Design patent infringement is fairly easy to catch because the designs are readily visible. Utility patent violations are harder to detect. Their workings can be buried inside much larger mechanisms or processes. That's particularly true of patented software, whose coding can be usurped and hidden deep inside other programs that consist of millions of lines of code.

Business process patents can be equally hard to catch when they are part of internal operations. Suspicions are aroused for processes aimed at consumers, such as how purchases are carried out on a Web site, when they function similarly in different settings.

What do you think of this? *Bilski and business process patents*

The debate over the soundness of the notion that business processes are patentable reached a turning point with the *Bilski v. Kappos* case that worked its way to the Supreme Court. In 2010, the Court overturned lower court decisions by holding that it was not sufficient to declare a business process patentable solely because it was tied to "a particular machine or apparatus" or that "it transforms a particular article into a different state or thing." A patent could be also denied if it infringed on a pre-existing process idea, even an abstract one.

At first blush, the decisions appeared to make business process patents much more difficult to acquire, but the Court's conclusions were by no means unanimous. Even the consenting opinions of the majority had some areas of difference, as did the dissenting opinions. So the issue of business methods patentability is still with us.

Taking unfair advantage

One tactic to achieve a competitive advantage and discourage others from even attempting to create a competing product is the *patent thicket*—dense combinations of many interrelated and overlapping patents held by one firm. The idea is that potential competitors or innovators have to work through the thicket to make sure that what they are developing won't be in violation of one or more of the patents it contains. It's perfectly legal, but is it ethical?

Thickets can comprise thousands of individual patents, making the task all but impossible, or at least extremely time consuming and expensive. That's cited by those who oppose thicket creation as being an anti-competitive anti-innovation tactic whose sole purpose is simply to keep rivals at bay. Supporters claim that the barriers thickets create induce the holders to innovate even more so as to strengthen their positions. One result of the practice is that many hopeful innovators proceed without an intensive investigation, taking a risk that they will not be in violation or that if they are, it won't be discovered.

A much more dubious practice is the business model followed by so-called *patent trolls*. These individuals and companies build large portfolios of patents, usually by buying them at fire sale prices from bankrupt businesses or struggling cash poor firms. They have no intention of using the patents to market products. Instead, they search for infringing use and then seek royalties, licensing fees, or lump sum payments that typically are out of proportion to the value of the patent for the product or service involved.

This malicious practice succeeds especially well when used against small companies that don't have the wherewithal to defend against a lawsuit that can easily run into the millions of dollars. So the threat of a suit makes it likely that they will settle, even if the defendant company is not actually infringing. What we have here is an insidious form of blackmail. How's that for business ethics?

Legislative difficulty   Odious as it may be, patent infringement pursuit by trolls is not illegal. Dealing with it legislatively is deceptively complicated.

One approach suggested to short-circuit troll lawsuits is to require that a patent be put to use within a given time period before it can become the basis for an infringement claim. That sounds like a good idea until you realize that it's not unusual for considerable time to pass before an inventor can bring a product or process to market, so a time limit could penalize legitimate developers. Moreover, even if a patent is never brought to market, why should the actual creator have to face diminishment of property rights, including the right to sell the patent, even to a troll?

A more practical line of attack is disallowing issuance of an immediate permanent injunction that prevents any use of an infringing patent out of hand. That idea grew out of a case that began when MercExchange (ME) sued eBay for using the "buy it now" option, which ME claimed infringed on one of the patents in its portfolio. In 2003 a Virginia court agreed, so ME sought an injunction against eBay. The request was denied by the District Court, but on further review, the US Court of Appeals overruled the denial and granted the injunction.

The case eventually reached the Supreme Court, which found that an injunction was too drastic a step to take unless standard equity principles were considered first—whether an injunction was needed to prevent significant irreversible damage to the patent holder and whether some form of financial compensation would resolve the issue without an injunction. The Court also noted that whether or not the plaintiff had put the patent to use was irrelevant in deciding if granting protection from infringement was appropriate.[41]

Patent cache—the new battle armor and armament   Patents for the right technology have always been valuable items, both for the protection they provide against infringement and for the potential they represent for reaping licensing fees. In the wireless world, another dimension has been added—survival. Companies like HTC, LG, Nokia, and Samsung that make cell phones depend on operating systems from companies like Google (Android), Apple (iOS), and Microsoft (Windows phone). Those companies in turn depend on the equipment manufacturers choosing their systems.

Because of this interlocking relationship, lawsuits for operating system patent infringement hit equipment manufacturers as well, who can end up having to pay licensing fees to one company based on a suit against another. A case in point is HTC, which pays a fee to Microsoft for every Android-equipped phone it sells. That comes back to Google, potentially hurting its competitiveness and attractiveness to phone makers.

In light of a surge in lawsuits and demands for licensing fees, corporations take to building huge caches of patents, both as protection against lawsuits and as a source of suits against competitors. In December 2010, CPTN Holdings, a company organized by Microsoft that includes Apple, Oracle, and EMC, bought 882 patents from a dying Novell for $450 million after it was acquired by Attachmate. In July 2011, a consortium led by Apple and Microsoft, along with Sony, Research in Motion, Ericsson, and EMC, paid $4.5 billion for a collection of 6,000 patents held by the bankrupt Nortel Networks.

Both these purchases were considered by Google to be collusive, preparatory to an attack on Android. In July 2011, Google bought 1,000 patents from IBM for an undisclosed amount. Then in August 2011, Google announced a $12.5 billion offer to buy Motorola Mobility,[42] primarily to gain its store of about 17,000 patents and possibly another 7,500 still under review by the Patent and Trademark Office. That puts Google in a much stronger position to defend against lawsuits and at the same time protect equipment manufacturers using Android as well.

In bulking up their patent portfolios, the major players seek protection against lawsuits and the possibility of finding a few patents that let them go back at their competitors. That's an understandable business practice, but at what cost to the industry and the nation? An issue: what happens to innovation and competition when patents become armor and armament rather than incentives for producing new products?

Patent caches or not, lawsuits and countersuits parade on. Apple is threatening Samsung and Motorola, claiming their Android phones infringe on Apple patents. Oracle is suing Google for $6 billion for Android violations of their patents. Microsoft joined the parade as well, as did Nokia, which recently settled a dispute with Apple. HTC recently lost an infringement case instituted by Apple. Undoubtedly more are in the works. So companies have moved competing in the marketplace with their products and services to lawsuits. Is that good for innovation? for creativity? for the revenues of lawyers and law firms?

Trademark—a little history

Simple quasi-trademarks began thousands of years ago as physical marks to identify ownership and distinguish one maker of goods from another. They were not full trademarks in today's sense of the word because there was no formal system to ensure that the marks were unique or afforded specific protections. The concept of a trademark serving to identify a specific company as the source of a product and owner of a brand remains.

Early in US history, similar to copyright and patents, intellectual property represented by trademarks was legally protected only by state acts. Informal trademarks were already in widespread use, though—brands seared into the hides of cattle as proof that they belonged to a particular ranch; names and logos impressed onto furniture to indicate craftsmen or manufacturers.

Partially formalized in 1788 for the makers of sailcloth, it wasn't until 1870 that general federal trademark legislation was passed, under the same congressional authority that underlied copyright and patent legislation. In 1881, the trademark law was rewritten, this time based on the commerce clause of the Constitution.

> The Congress shall have Power . . . To regulate Commerce with foreign Nations, and among the several States, and with the Indian Tribes. [43]

The concern then was with international trade. Within the US, protection was left to the individual states. That changed in 1905 with the passage of the *Trademark Act* that extended federal trademark protection to interstate commerce.

Several minor modifications were enacted over the years. Then in 1946 legislation that became the basis for current trademark law was passed. Called the *Lanham Act*,[44] one of its many provisions helped settle disputes over prior state-granted trademarks.

This Act also was modified over time. Among the changes were those in recognition of advancing technologies and their previous lack of trademark protection. So not only were words and logos trademarkable, but also colors, sounds, and aromas.

For example, the United Parcel Service (UPS) has trademarked "See What Brown Can Do for You" in addition to other similar phrases that use the word brown, including the word itself. These have now become closely associated with the UPS, creating a specific brand identity.[45] That has enabled them to sue other companies for trademark dilution for using "brown" in URLs, and service and product references.

Current law gives trademark holders exclusive ownership rights forever, renewable every ten years for a fee, provided the trademark is used. If not, ownership is lost and another firm can claim the mark.

Types of trademarks

There are four kinds of marks, categorized for convenience. All have the same protections and application requirements. When related to services, trademarks are called service marks. We use the term trademark to cover both.

- *Descriptive*—directly invokes the marketed goods or services. Examples: Park & Fly; Jiffy Lube; Brides Magazine; Apartments Unlimited; Roach Motel; Cream of Wheat; Play Doh.

Not every descriptive phrase is trademarkable. Shredded wheat and cornflakes are not, because they've been long-standing generic descriptions of cereal types. On the other hand, Kellog's has been granted trademarks for Frosted Mini Wheats and Frosted Flakes because years of widespread advertising associated them with particular products of the company, therefore part of their branding.

- *Suggestive*—indirectly refers to goods, services, or their quality. Examples: Juicy Fruit chewing gum; Meow Mix cat food; 7–11 convenience stores; Papermate pens; Seahorse boats; Fruit Loops cereal; Home Depot stores.
- *Arbitrary*—common words or names not related to actual products or services. Examples: Shell oil company; Lipton tea; Lifesavers candy, Maxwell House coffee. Acronyms also are considered arbitrary, even if they stand for actual names—examples: UPS; IBM; NBC; AMC. Colors also fit the category: UPS's brown; Owens Corning's pink; Coca Cola's red.
- *Fanciful*—made-up words that have no real meaning. Examples: Kleenex; Cheerios; Verizon; Magnavox; Exxon.

Corporate names are not automatically trademarked. To receive a trademark, an application has to be filed with the Patent and Trademark Office and be approved. If a name isn't trademarked, many businesses can use it or similar names. For example, there is more than one Napoli Pizzeria, Joe's Towing Service, and Rose's Florist Shop.

Trademark dilution   Another one of those simple yet controversial concepts that can get quite murky is *trademark dilution*. The idea is unambiguous—since a trademark confers valuable cachet and recognition, similar names, logos, color usage, and the like dilute that value by customers confusing one company with another. Aside from the value of the mark itself, such confusion can harm the image of the first-marked company, resulting in straightforward economic damage. That the companies are or are not competitors is often immaterial.

Considered from a consumer perspective, dilution is disallowed to protect them from a natural assumption that the reputation of one company is deserved by another. That confusion is why so-called knockoff goods, which carry the marks of well-known companies but are actually shoddy imitations, are popular with the imitators—they trade on the reputation of the business whose mark they use. Consumers who buy the merchandise may be so desirous of having the cachet of a "name brand" at a very low price that they are either too naive to realize they are being scammed or too uncaring. Either way, knockoff makers thrive.

First to trademark does not always mean being protected from usurpation, nor does lack of confusion mean dilution can't be successfully claimed. Often, dilution arguments have the most sway when one company is much more powerful and well known than another, even when there seems to be little to no chance of confounding the two and even when the smaller company existed before the larger one. On the other hand, many similar trademarks have been granted on just that non-confusion ground. For example: General Electric, General Motors, General Vision, and General Mills are not related companies, nor are Quaker Oats, Quaker Maid, Quaker Windows, and Quaker Chemicals.

The philosophy of trademarks differs from that of copyrights and patents. Trademarks focus exclusively on protection. Copyright and patent mechanisms also aim at encouraging innovation and spreading information. In the scheme of things, trademarks can represent enormous value, but promoting creativity is not within their scope. In terms of societal value, copyrights and patents come out ahead, but has applying what many consider to be outdated laws moved us from a practice that encourages creativity and innovation to one that discourages them?

And so—innovation stimulator or destroyer

What to do about intellectual property? It's hard to argue against the idea that creators of intellectual property should be entitled to benefit from their work if they so choose, to have an opportunity for, though not a guarantee of, recompense. As we've seen, historical precedents for protecting intellectual property abound. The noble effort to give IP creators protection via property rights has worked reasonably well over the years, but there are growing issues about the appropriateness of the laws given the advancements in IT, especially digital transmission and processing techniques.

Justified by the concept of incentivizing further work and formalized in legislation based on property rights, copyright and patent laws endeavor to create reasonable protections in the face of complex problems. Attempts to resolve these issues have led us to a place where the zeal to protect has so bottled up the process of creativity that innovation has declined. The balance has tipped too far towards exclusivity. After all, the idea of incentivizing is not just to reward the creator of some intellectual property, but to stimulate others to build upon it as well.

That said, striking a balance is not easy. Too much regulation, or too little, and innovation declines. Boiled down to their essentials, the issues in flux are: how much protection, under what circumstances, and with what enforcement methods?

Copyright—sensible and strange   You can copy anything if you get permission to do so, but to completely close off usage without consent doesn't make sense. That's why fair use rules were instituted. They recognize that allowing small amounts of attributed content without approval is a good idea, if judiciously applied. The significance of the copied segments is as important as how much was copied. If they are the heart of the work, even if that's a relatively small proportion of the whole, fair use will likely not apply. That these sensible conditions are met is not always easy to determine. After all, what qualifies as the heart of the work and how much is relatively small are, at base, judgment calls.

Disregarding any subjectivity is the oddity of music sampling, which we've found is treated differently. Taking the smallest snippets of music is infringement, regardless of significance. Fair use doesn't apply, it seems.

Software presents other difficulties. Is copying a line of code fair use if attributed? A great many lines of code to accomplish a particular function have converged over time, so common as to be considered standard programming practice. For all practical purposes they are in the public domain, though not necessarily legally so. If two different programs use significant amounts of the same such code, that could be the basis for a copyright infringement claim.

Copyright term length has grown from its original 14 years to a whopping 70 years past the life of individual holders. Does that term excessively limit others? Would holders have sufficient opportunity to capitalize on their work with shorter terms, say, for example, the 20-year term that patentees receive?

We've seen that there is copyrighted material lying moribund, unusable even though the copyright holders can't be found, or they are dead and have no heirs. Should there be some mandated requirement to find the missing persons or inheritors? What kind of search, for how long? Or should copyrights for such inaccessible works be revoked? So far, it's been easier to just let them lie.

Here is a small sample of actual copyright cases to ponder. You be the judge.

● In 1952, avant-garde composer John Cage released *4'33"*, a piece consisting of musicians not playing anything for four minutes and 33 seconds—that is, silence, except for some hardly noticeable random background sounds. In 2002, Mike Batt and his band *The Planets* produced an album with a track titled *A One Minute Silence*, attributed to Batt/Cage. The estate of John Cage sued for copyright infringement. Eventually Batt settled by paying the estate a six-digit sum.

Looking back at fair use, was one minute too much a part of the whole? Was the attribution insufficient? Could the Batt piece be considered parody? Could anyone determine which minute of the 4'33" was copied? Common sense would indicate this was a frivolous suit. So why did Batt settle? Perhaps it was cheaper than paying lawyers to mount a defense, and less time consuming as well.

● In 1962, Heloise Pinheiro was a sultry 17 year old from Ipanema, Brazil, who inspired Brazil's best-known song, *The Girl from Ipanema*. A 1965 version won the record of the year Grammy. Not long after, she opened a boutique by the same name. That caught the eye of the song composers' heirs, who sued her for copyright infringement.

Would there likely be any confusion between the song and the boutique? Would the operation of the boutique have any negative impact on the heirs' income from the song? Should Pinheiro have to change the name of the shop or pay the heirs for use of the name? Should she sue the heirs for harassment? Another seemingly frivolous suit, this one went nowhere.

- In 2003, boxer Mike Tyson had his face tattooed by Victor Whitmill, who also designed the tattoo. In 2011, Warner Bros. released the movie *The Hangover Part 2*, in which one of the characters had his face tattooed with a design mimicking Tyson's. Warner Bros. was sued by Whitmill for copyright infringement.

The movie was classified as a comedy. Does that make use of the design a parody? Was irreparable harm done to Whitmill or did the exposure he received from the tattoo appearing in the movie and the publicity from the lawsuit result in just the opposite? Since Tyson paid for the design and the tattoo, does that make him the actual copyright holder, or at least a shared owner? If Whitmill applies the same tattoo to others, could he be sued by Tyson? If Tyson appears in an ad for which he is paid and in which the tattoo is clearly visible, could he be sued by Whitmill? The outcome of Whitmill's lawsuit: it was settled but the agreement wasn't disclosed.

Patents follow a similar path   Patent evolution tends to be a little more consistent than that of copyright, with fewer twists and turns. There are notable exceptions, though. In particular, the soundness of granting business and software patents has been called to question, as has the affect of biologic and pharmaceutical patents on research and marketing.

Subjective judgment presents the same kind of dilemma for patents as for copyrights. Evidence of usefulness, novelty, and inventiveness are required to receive a product patent, but how useful, how novel, and how inventive are issues. Then too, though the quid pro quo for patent protection—full disclosure of details—is presumed to give others information they need to improve upon the work while the holder benefits from exclusivity, that the wait before details are released may be too long is yet another issue.

Consider how the rules have operated for major pharmaceutical companies. They spend a great deal of time and money developing new medicines. If proven efficacious, they can be patented. During 20 years of patent life, drug prices are very high—after all, the companies have to recoup their development costs, overcome expenses for research that didn't lead to marketable drugs, make a profit, and fund further research. To that end, they boost sales of new drugs by heavily marketing them to consumers, especially through the media, even though they are not purchasable without a prescription. Of course, that adds to the price too.

Still and all, reasonable arguments can be made for much of that practice, until we realize that patent incentives can work the wrong way—yet another unintended consequence. How do pharmaceutical companies decide which drugs to pursue? As profit-making businesses, they would be wise to seek products most likely to become profitable after the shortest development time. That's why research often aims less at the more intractable diseases and small market maladies. Potential societal benefit plays a lesser role.

As the 20-year end approaches, generics makers gear up production. They can sell copies of the drugs at considerably lower prices because their development and marketing costs are a good deal less. A great many customers believe generics are as effective as brand name drugs and they do have to meet content and formulation requirements, so sales of the brand names decline significantly. In reaction, the original makers take to the offense by altering their products just enough to obtain new patents—novelty achieved—and then market the new formulations heavily, whether or not they are any more efficacious than the ones they replace.

Looking at patent litigation in general, we find rulings that appear reasonable and others counterintuitive. Once again, you be the judge as you ponder these examples.

- In 2003, Beverly Hills patent lawyer Frank Weyer and partner Troy Javaher, through their newly formed *Nizza Group*, received a patent for assigning email addresses to a group by replacing one dot in the name domain URL with an @. So for a URL of www.john.smith.com, the e-mail address would be john@smith.com. No strangers to lawsuits, they promptly sued two leading domain name registrars *Network Solutions* and *Register.com* for patent infringement, asking for money and an injunction against issuance of name domain URLs.

Did the examiners who approved the patent understand whether or not Nizza Group's method had the requisite novelty or improvement over existing methods? Did they realize that, in essence, the patent application had simply described an existing naming convention? Should any form of URLs and email addresses be patentable in the first place? Outcome: the case was dismissed.

- In 2000, David Roth had a brainstorm—cereal, beloved by generations, could be turned into a profitable business. He and a partner opened *Cereality* at the University of Arizona campus, an eatery serving concoctions of branded cereals and toppings. It was a big success. A little later, Rocco Monteleone came up with the same idea and opened the *Cereal Bowl* near the University of Florida. Afterwards, he learned of Cereality, but figured it didn't matter. There are vast numbers of eateries selling pretty much the same things as each other, so why not another cereal shop? Monteleone found out why not, in a letter from Cereality's attorney. The company had applied for a business methods patent, the method being serving mixed cereals with toppings and liquids.

Digital processing was the impetus behind modifying the law to allow software and business process patents, the latter almost always comprising software. That had the unintended consequence of opening the floodgates to patent applications and infringement claims for all sorts of business processes, software based or not.

Cereality's patent had nothing to do with software. Is there anything unique about their method, or is it something that millions of people do every day in their kitchens? Is there any element of novelty or anything not obvious to the average person? Does the patent mean that no other eatery could serve cereal without permission from or payment to Cereality? Would such a patent open the door to patents for serving burgers with toppings on a variety of buns, different preparations of eggs, various sandwiches? Outcome: the patent was denied.

● In 2001, Australian patent attorney John Keogh received a patent under the newly passed Australian *Innovation Patent* system, for a "circular transportation facilitation device."

Australia's innovation patent system was well intentioned. It was to be an inexpensive, simple, fast means for small businesses to obtain patents without needing lawyers and to short-circuit the standard lengthy evaluation process. The idea was simply to stimulate innovation, and so, time-consuming and expensive demonstration of a significant advance took a back seat. Outcome: Keogh received a patent for what we know more commonly as the wheel.

Good intentions not withstanding, this happenstance was a classic example of unintended consequences. Its provisions were too loose, the fatal flaw that Keogh seized upon. It's doubtful that he would have prevailed had he instituted lawsuits based on his patent, but that was not his goal. He merely wanted to point out the absurdity of the presumably well-intentioned law.

Trademarks—not quite as clear-cut as they seem   Trademarks are the most straightforward and least controversial of intellectual properties, but they are not exempt from incongruities and disputes. Trademark dilution claims, in particular, have been contentious.

Made easier by passage of the *Trademark Dilution Revision Act of 2006*, injunctions can be sought if dilution "by blurring or tarnishment" is likely whether or not there is any actual confusion as to trademark ownership or any economic injury to the trademark holder.[46] So even a minimally similar logo, shade of color, or wording is sufficient to be the basis for a dilution claim. Does that subvert the purported intentions of trademarks?

Let's look at some actual claims. What would you conclude?

● The Girl from Ipanema copyright case was not the end of the story. Not to be left out, Astrud Gilberto, the woman who sang the Grammy winning recording, claimed that she had become known as the Girl from Ipanema because of the Grammy and her frequent performances of the song. She claimed that she and the song were inextricably intertwined, to the extent that it had become her trademark. Gilberto sued Frito-Lay for trademark infringement for using the recording in an ad without her permission, which implied her endorsement of the product.[47]

Would the public assume that the ad carried Gilberto's endorsement? Does the fact that she didn't write the song or the lyrics and didn't produce the original recording mean she can't claim trademark rights? Did the ad campaign provide Frito-Lay with economic benefit at the expense of Gilberto? Can trademark rights accrue to someone who hadn't applied for the mark? Outcome: the suit was dismissed.

● Early in 2003, *Hooters of America*, whose claim to fame rests primarily on food service by well-endowed waitresses clad in tight-fitting scoop neck tank tops and short shorts, sued *WingHouse* in a Florida US District Court, claiming "trade dress infringement and dilution in contravention of the Lanham Act."[48] WingHouse, they said, infringed on Hooters' costumes and design features of their establishments, such as surfboards, parchment menus, beach themes, and the like, and by doing so, diluted the value of the Hooters brand.

Infringement requires that "1) its trade dress is inherently distinctive or has acquired secondary meaning, 2) its trade dress is primarily non-functional, and 3) the defendant's trade dress is confusingly similar."[49] Was that the case? Were Hooters' design elements distinctive or were tank tops and shorts a common clothing style? How about the décor? Outcome: the claim was denied. Hooters appealed, but in 2006 the US Appeals Court for the 11th District upheld the original ruling.

- In 2000, Samantha Lundberg, née Buck, known as Sam Buck to folks in the small town of her birth, Astoria, Oregon, opened a coffee shop under her run-together maiden name, Sambucks. A year later she received a cease and desist letter from Starbucks offering her $500 to change the name. She countered with a demand for $60,000. Starbucks pursued the issue by suing for trademark dilution.

There is no question that the design of Sambucks' shop, logo, color scheme, and even coffee, were all significantly different from Starbucks. The chance of anyone assuming that the shop was affiliated with Starbucks was nil. Was it foolish for Starbucks to pursue the case?

Precedent is the key. If Starbucks declined to pursue one potentially diluting use, as minor as it may have been, other and bolder attempts would likely follow. The more attempts, the more costly and time consuming it becomes to defend against them. So it's better to be aggressive early on—but is there no point at which a dilution claim is frivolous? Outcome: Starbucks won.

- In 2005, *Virgin Group*, parent company of a multitude of similarly named businesses—Virgin Airlines, Virgin Racing, Virgin Mobile, and Virgin Megastores to name but a few—in actions similar to Starbucks, sent cease and desist letters to a number of enterprises that used the word virgin in their names or URLs. Except for Jason Yang, sole proprietor of *Virgin Threads*, they capitulated. Yang did not and was sued for trademark dilution.

Yang's business, selling fashions of emerging independent designers via his Web site, virginthreads.com, had about $105,000 in sales the year before the suit; Virgin Group's total, $8.1 billion. Was Yang's operation a threat to the Group? Did the business deprive the Group of income or damage its reputation? Was there any confusion about ownership on the part of the public? Is it sensible to allow trademark protection for marks based on a common word like virgin? Did any of those issues matter? Outcome: Virgin Group won. Virgin Threads changed its name to Stars and Infinite Darkness. Its Web site is still under construction.

Did the fact that Starbucks and the Virgin Group are vastly larger corporations than Sambucks and Virgin Threads have any bearing on the outcomes? It did, for two reasons: small businesses can ill afford to fight lawsuits against large firms; small firms can benefit from being mistakenly associated with large business, but not the other way around—only large firms experience dilution in such very uneven comparisons.

We could say that very small businesses are unlikely to invoke mistaken associations and so could be needlessly penalized by dilution claims such as we've seen. That may be, but is it clear in any situation whether a small business just happened on a name or innocently infringed, or was actually trying to play off of a trademark's popularity or get paid to change its name?

When the names in question are fanciful, like Altria or Xerox, there's no doubt that incorporating them in a business name or URL is trademark infringement and dilution. Motivation is not so obvious when common words are involved, or the business name is the person's name. That's the dilemma. In courts the balance often tilts to the big firm.

Think about it this way:

> . . . big money is at stake. Aspirin was once a brand name. [So were] thermos, trampoline, cube steak, cellophane, elevator and escalator. The inventor[s] of those products lost their valuable exclusives on those names because they fell into common use.[50]

On the other hand, there's a big difference between being careless or inattentive in protecting a trademark and going after every possible incident.

It all comes down to this   One person's intellectual property rights is another's inanity. We've seen both reasonable and bizarre protections upheld, and both reasonable and bizarre protections dismissed. We've found that the overwhelming resources available to large-scale corporations put them in the driver's seat, giving them substantial control over innovative practices and discouraging others from pursuing their own ideas—incentivization turned on its head.

So you can think about it this way: copyrights, patents, and trademarks are meant to protect innovators and creators of intellectual property, whether people or corporations, from usurpation of their ideas. They are not meant to give deep pocket companies a legal means of intimidating people and small companies by threats of legal action that they can ill afford to defend, so as to either damp down potential competition or reap undue rewards. On the other hand, they also are not meant to enable small aggressive companies from benefiting from false association with better known firms or profiting from illegal use of rightful intellectual property.

Laws regarding intellectual property protection have worked fairly well over the years, but problems are growing the more digital technology is part of the picture. Aside from the details of length of protection and fair use, the balance between appropriate protection and reasonable use needs adjusting.

Any hope of finding a new balance has to take into account that, despite the fact that protections all are based on the concept of property ownership, there is neither a strong commonality among the different types of intellectual property nor between ephemeral digital intellectual property and real property. Laws addressing intellectual property have evolved as modifications of real property concepts. That was workable before the digital revolution took hold. Now they can't rely on parallels between the two.

Notes

1   Tim Berners-Lee as quoted in: http://www.brainyquote.com/quotes/quotes/t/timberners373107.html
2   Lawrence Lessig interview: http://openp2p.com/pub/a/p2p/2001/01/30/lessig.html?page=3
3   US Constitution: Article 1—The Legislative Branch; Section 8—Powers of Congress, clause 8: http://www.house.gov/house/Constitution/Constitution.html
4   Facts can't be copyrighted but how they are described or assessed can be.
5   http://www.copyright.gov
6   For the full text of all US federal copyright laws see http://law.copyrightdata.com/
7   Lawrence Lessig interview: http://openp2p.com/pub/a/p2p/2001/01/30/lessig.html?page=1
8   Shugart Technologies, founded in 1979, soon changed its name to Seagate Technologies.
9   Full text of the PCA Act (Public Law 97–180) is at: http://law.copyrightdata.com/public_laws.php
10 Full text of the case is at: http://www.law.cornell.edu/copyright/cases/464_US_417.htm
11 That assertion was based on the retail value of the copied material, assuming that every copy would have been a sale—quite unlikely. Any way you look at it, though, there was a lot of copying going on.
12 *Copyright Infringement—Fourth Element—Commercial Advantage or Private Financial Gain*: http://www.justice.gov/usao/eousa/foia_reading_room/usam/title9/crm01851.htm
13 *US v. David LaMacchia*, Crim. A. No. 94–10092-RGS, US District Court, D. MA, December 28, 1994: http://scholar.google.com/scholar_case?case=13104224228509781852&q=U.S.+v.+LaMacchia,+871&hl=en&as_sdt=2,33&as_vis=1
14 *Dowling v. United States*, 473 U.S. 207 (1985).
15 Full text of the NET Act is at: http://www.techlawjournal.com/courts/eldritch/pl105–147.htm
16 Full text of the Treaty is at: http://www.wipo.int/treaties/en/ip/wct/trtdocs_wo033.html. A list of the 89 contracting nations through 2010, with contract signing dates, is at: http://www.wipo.int/treaties/en/ShowResults.jsp?lang=en&treaty_id=16
17 Full text of the DMCA is at: http://w2.eff.org/IP/DMCA/hr2281_dmca_law_19981020_pl105–304.html
18 Although Napster's sharing model could work for any kind of file, music was what it was all about.
19 Two major opponents of DRM are the Electronic Frontier Foundation (EFF): http://www.eff.org/ and the Electronic Privacy Information Center (EPIC): http://epic.org
20 The BitTorrent company Web site homepage is at: http://www.bittorrent.com/
21 http://www.gnu.org/licenses/gpl.html
22 http://creativecommons.org/
23 A dongle is a small device that needs to be connected to a computer for protected content to be accessed.
24 For more information, see Ben Sisario, "To Slow Piracy, Internet Providers Ready Penalties," *New York Times*, July 7, 2011: http://www.nytimes.com/2011/07/08/technology/to-slow-piracy-internet-providers-ready-penalties.html?_r=1&scp=2&sq=copyright%20alert%20system&st=cse
25 http://radiohead.com/
26 http://www.washingtonpost.com/wp-dyn/content/article/2007/10/10/AR2007101002442.html
27 http://www.apple.com/
28 http://www.jamendo.com/en/
29 http://www.pandora.com
30 http://www.rhapsody.com/
31 http://magnatune.com/
32 http://www.amazon.com/gp/feature.html?docId=1000545341&tag=amistr-20
33 http://songza.com/
34 Full text of the 1790 Act is at: http://ipmall.info/hosted_resources/lipa/patents/Patent_Act_of_1790.pdf
35 Full text of the 1793 Act is at: http://ipmall.info/hosted_resources/lipa/patents/Patent_Act_of_1793.pdf
36 Full text of the 1836 Act is at: http://ipmall.info/hosted_resources/lipa/patents/Patent_Act_of_1836.pdf
37 *Hotchkiss v. Greenwood*, 52 U. S. 248: http://supreme.justia.com/us/52/248/
38 See: http://www.wipo.int/pct/en/texts/articles/atoc.htm

39 http://www.patentoffice.com

40 Miguel de Icaza (1972–), software programmer, quoted in: http://www.brainyquote.com/quotes/keywords/patent.html

41 US Supreme Court No. 05–130, Ebay Inc. et al. V. Mercexchange, L. L. C., Certiorari to the United States Court of Appeals for the Federal Circuit, No. 05–130. Argued March 29, 2006—Decided May 15, 2006: http://www.supremecourt.gov/opinions/05pdf/05–130.pdf

42 January 2011, Motorola split into two companies—Motorola Mobility, which makes mobile phones, set top boxes, and other consumer products; Motorola Solutions, which sells equipment to businesses.

43 US Constitution, Article 1 Section 8 Clause 3 (commerce clause): http://www.house.gov/house/Constitution/Constitution.html

44 Details of the Lanham Act, actually title 15, chapter 22 of the US Code of Laws, is at: http://www.bitlaw.com/source/15usc/

45 For a list of UPS trademarks, see: http://ups.org/media/en/trademarks.pdf

46 The act originated as H.R. 683 introduced in 2005. In 2006 it was passed and signed into law: http://www.loeb.com/trademarkdilutionrevisionact2006/

47 http://openjurist.org/251/f3d/56/astrud-oliveira-v-frito-lay-inc-

48 http://howappealing.law.com/HootersVsWingHouse.pdf

49 *Ambrit, Inc. v. Kraft,Inc.,* 812 F.2d 1531, 1535 (11th Cir. 1986), *cert. denied,* 481 U.S. 1041 (1987).

50 http://abcnews.go.com/2020/GiveMeABreak/story?id=1390867

6

You, me, and everyone else—alone together and vice versa

It has never been easier to be as influential as you can be today. Information is cheap. Information is easier to produce. And if you have a quality message, it's never been cheaper to get out.[1]

Social media can be an enabler and an accelerator of existing core capabilities, values, attributes and plans. It can even be a catalyst for change. But it can't magically create what doesn't exist.[2]

Relationships

Humans are social animals. Personal relationships of all sorts are part of life. Whether in formal structures like family or business, informal friendships, casual or merely nodding acquaintances, with extremely rare exceptions we don't want to live our lives in isolation.

Relationship structures can be simple or complex, direct or indirect, close or distant. The structures of many people can be intertwined, leading to quite complex configurations. We call those relationship structures *networks*, a term that connotes arrangements.

Whatever kind of relationships we're talking about, communication and contact are required to build, maintain, expand, reduce, and change them. For a couple of centuries before the Internet, keeping in touch with a distant acquaintance, friend, or relative meant writing a letter or making a landline telephone call. The former was very slow, the latter could be quite expensive and you might not reach the person at all.

It was worse for group communication and businesses. Mailing the same message to a number of people meant repeating the process for each one. If a lot of information needed to be sent with each letter, that was even more tedious and expensive.

Over time the nature of communication and contact changed, most notably in the recent past with the advent of digital technology. The transition from analog to digital, from in-person to virtual, from local to global, brought about a remarkable transformation in our ability to communicate and in many respects the meaning of relationships.

The issues involved are many, varied, and complex, too much so to be fully explored in one chapter. So we look at major areas of change that have come about as people adopted new communications technologies and applications. In that milieu, social networks and online communities are at the forefront as instruments of change.

As we'll see, though, their adoption and uses have extended well beyond what was envisioned at the outset.

Some definitions

Before we explore the impacts of social engagement, we offer these definitions:

- *social*—involving people
- *network*—organized interconnections descriptive of technology or relationships
- *media*—means of providing information, usually referring to mass communications
- *networking*—the act of using networks or creating connections
- *community*—a group of people who have something, though rarely everything, in common—locality, culture, history, government, religion, occupation, interests
- *social network*—refers to two different things: an interconnected communications infrastructure; a depiction of intertwined social relationships. Context usually suffices to indicate the intended reference
- *social media*—applications like Facebook, Twitter, and foursquare that provide services for constructing and using personalized social networks
- *social networking*—the act of utilizing social media to build connections and communicate with them and others
- *online community*—the online version of a community. Members are not constrained by geography or time. *Virtual community* is a synonym

A little history

As a technology, a network is a communications framework, an information transmission system, a map of connections. Add "social" and the focus changes to a means for people to interact within a communications system. The actual relationships that a social network supports are a separate matter.

When many of us think of social networking now, what comes to mind are sites like Facebook, LinkedIn, and Twitter, currently the most popular among a large number of computer-based enabling systems, most of which ride on the Internet. One of the more important and impactful phenomena of the IT age so far, they are only the latest in a long line of broadly defined social networks that have their beginnings at the dawn of civilization.

Social interaction is a basic human instinct that people have always sought. Tribal imperatives, friendships, sharers of common interests, news seekers, news relayers, and the like created webs of connections among the earliest civilizations by simple conversation. As communication technology improved, so did the opportunities for broadening those webs.

Flash forward to applications of electricity. The growing speed, reach, and capabilities of communication systems prompted more frequent and more rapid interaction among people. Networks of business and personal contacts expanded greatly.

These developments and similar ones, dramatic as they were in their time, grew slowly in size, scope, and span over many years. The communication technologies they rode on—including the telegraph, telephone, broadcast radio and television, of which only telephones provided real-time two-way communication—were certainly impressive in their day. Yet they did not deliver nearly the impetus provided by widespread use of computers, mobile systems, and that über network of networks, the Internet—all computer mediated means of digital communication.

Bulletin boards—a precursor   In the late 1970s, decades before the Internet was readily available to the general public, hosted bulletin board systems (BBSs) came into being. These were electronic analogs of physical bulletin boards and played much the same role—notices of items for sale, jobs sought and available, community activities, and the like. Special interest groups found BBSs to be an efficient way to point members to events. Businesses used them to spread company-related information, often by posting start-of-day announcements.

The functionality of digital computer-based BBSs was soon discovered. People readily took to treating them as venues for promoting individual viewpoints, providing helpful advice ad hoc or in response to questions, disseminating business directives, and transferring files. Illegal distribution of copyrighted software wasn't far behind.

BBSs created virtual communities—online societies of a sort—illustrating a major difference between virtual and physical communities: in the former, participants don't have to and may never actually meet in person, nor do they need to be in any particular physical proximity to one another.

The first BBS was created by Ward Christianson and Randy Seuss in Chicago, launched in February, 1978.[3] Called the CBBS (Computerized BBS), it initially ran on a single computer connected to one modem.[4] That meant that only one user at a time, mainly people in the IT field, could access the system. As the utility of BBSs became evident, installations grew at an increasing pace, shifting from the sole province of techies to a generally useful and intriguing communication modality.

Ric Manning, writing in 1986, said,

> Estimates of the number of bulletin boards in operation run upwards of 4,000, but nobody knows for sure because turning a computer into a BBS is like plugging in a telephone answering machine—a simple matter of personal communication choice.[5]

The estimates included a variety of modalities: some BBSs were private; some run mostly by hobbyists were free to the public; others were subscription based; and still others were operated by companies for their employees and customers.

BBSs depended on the existing analog telephone system. Participating members using computers equipped with modems connected to a host computer running BBS software by dialing a telephone number. The software also provided tools for system operators to manage the boards, organizing, distributing, and storing messages.

Fidonet, expanding reach   By 1984, IBM and Apple were on the road to becoming significant players in the personal computer arena and the Internet was starting to develop. So it would seem that for BBSs, personal computers and the Internet were a natural pairing, the latter being much more suited to digital communications than the analog telephone system. But there were growing pains.

Compatibility was a problem. The Internet ran on the UNIX operating system, quite different from those used on IBM and Apple machines. Increasingly popular email was served principally by the BBSs and could not readily travel over the Internet from those computers. Then too, one BBS could not easily communicate with another.

Fidonet,[6] a telephone/modem-based store-and-forward point-to-point system for email and file exchange, addressed the BBS intercommunication issue. Developed by Tom Jennings, it went live in 1984. BBSs that ran Fidonet software could pass messages among themselves. That led to its growing popularity through the early1990s, but declined as the Internet and various Web applications gained ground, even though some versions of Fidonet could use the Internet for message transport.[7]

In a parallel development, online commercial organizations provided their own messaging services and acted as Internet portals as well. Their business models were based on usage fees for service, or as subscriptions with pay scales based on volume and connection speeds.

The interplay between growing personal computer ownership, connectivity options, and Internet capability consigned BBSs to a very small niche. More robust and flexible applications took center stage.

Next up, social networks   Internet development and growth simplified creating communications platforms for special interest groups, user groups, and bulletin boards, as well as email and instant messaging for the general population. As Internet and Web penetration increased, social networks and social media took advantage, starting slowly and then leaping into prominence. The list of aborted attempts, sites that had some success but eventually failed, and those that have so far succeeded, is quite large. Here are some key players.

In the 1990s, several applications hinted at the coming of social networks. *AIM* (AOL Instant Messenger) began in 1996 as a Buddy List for AOL members to contact each other.[8] *ICQ* debuted the same year, a similar service with a big difference—it was available to anyone with an Internet connection.[9] As a flexible free service, it quickly gained enough popularity to threaten AOL. Two years later, AOL bought it to protect its own market. A few other sites popped up around the same time, some earlier than AIM and some a little later. They targeted particular groups of people—classmates, professionals, relationships, and special interest groups that had been using bulletin boards—but few caught on to a significant extent.

An exception was SixDegrees.com, launched in 1997. It was unique in that it included not just lists of contacts and friends but also profiles, postings, and the ability to invite others to join. Those are the kinds of features we've come to expect from social networks and so, SixDegrees was credited with being the first. It quickly grew to prominence, gaining millions of users, but succumbed because of an unsustainable business model. Just three years after launch it closed. Still, it was the precursor of many sites to come, among which were Friendster, MySpace, and Facebook.

Friendster, launched in 2002, was the first social network site after SixDegrees to achieve wide success. Members could form networks of invited friends who could link to each other's networks, creating interlocking webs. It attracted millions of users in a few short months. By 2008 it had over 115,000,000 members and had expanded to other countries. It became a metaphor for Facebook, which together with MySpace ultimately did it in.

MySpace opened to the public in 2003, founded by Chris DeWolf and Tom Anderson who, somewhat ironically, were Friendster members. MySpace added features not prominent in other applications of the time, such as shopping, blogs, and personal pages on which members could post whatever they wanted to. It became a second metaphor for Facebook.

MySpace's considerable openness attracted millions of members, but also caused unexpected problems as sexual predators came to the site. In response, some security restrictions were added, but that cut down on its openness, one of the more popular features of the site. That illustrates the typical dilemma of action/response tradeoffs—where to draw the line between open access and protection.

Despite its problems, MySpace drew many users away from Friendster. It also was able to keep Facebook at bay for some time, but not forever.

Facebook started in 2004 as a social network site for Harvard students. It gradually was opened to other universities, then high schools, and finally to anyone at least 13 years old. Combining the features of Friendster, MySpace, and some other sites, it attempted to avoid the openness problems MySpace faced by adopting a long list of user policies that members had to agree to before being able to join. Adding other features, Facebook began to grow rapidly. It passed the 200 million user mark in 2008, 300 million in 2009, 500 million in 2010, and 800 million in 2011. It is still growing, though there is some indication that it is reaching saturation and a growing number of members, disenchanted, are leaving.

Controversy has not escaped Facebook. Members found that the site's default settings left their personal information readily available. Privacy issues became so contentious that Facebook took pains to simplify their arcane privacy setting routines. Still, members concerned about privacy and security issues also are beginning to leave the site. On another front, disputes among the founders and former partners led to lawsuits that went on for some years before being settled. Much of the controversy has died down and, at least for now, Facebook journeys on.

Two years after Facebook's launch, Twitter debuted. It began as a microblogging texting service. As it developed, search and group formation via hashtags were added. Twitpic appeared in 2008, an add-on application supported by Twitter clients. With it, users can post photos through Twitter in real time.

By 2011 there were over 200 million Twitter users worldwide. Lately, the younger generations have begun to favor Twitter over blogging. As we'll see, Twitter and Facebook have grown beyond benign socialization to become important tools for organizing and coordinating demonstrations, thereby playing key roles in supporting societal change.

Aspects of socializing

Socializing was harder in the pre-Internet days, or so it seems. You actually had to meet people face to face. Making friends took some time and effort. Now it's button clicks.

That raises a number of issues. Is friends the wrong word to describe these relationships? Has anything changed besides the manner of contact? Does socializing via chip-based devices mean social networks are isolating or are they simply a different way to be social? Does face to face matter any more? Those are the outlines of the debate. The answers are not yes or no—they are "it depends."

We're on the verge of having two generations of people who haven't known life before computers, the Internet, and mobile communications. Online social networking comes to them naturally. They have interwoven collections of friends, however defined, that are much larger than was typical when face to face was primary. Just as social networks have lifted geographical limitations of casual relationships, so have they become an easy way to keep in touch with family, close friends, and each other even when scattered across the globe.

Social networks are not just for teens, as evidenced by millions of registered users well past youth. While the youngest generations took to it as a matter of course, others of us had to come around to the idea. Either way, most began rather naively, with loss of privacy hardly noticed. Consequences ranged from embarrassment to being turned down for or fired from jobs. Readily available personal information also made it easier for Internet stalkers, scam artists, and bullies.

Social media are an apt platform for bullying in the digital era, yet another unintended consequence. So-called cyberbullying—the antithesis of socializing—turns social networking on its head. Tormenters often are anonymous, sending and posting hurtful or malicious diatribes, gossip, photos, and other hateful messages intended to embarrass or denigrate their victims. Some such attacks have had tragic consequences.

What do you think of this? *Cyberbullying's extreme outcome*

Although most cyberbullying is from teen to teen, adults get into the act too. In 2006, Lori Drew, the mother of a former girlfriend of 13-year-old Megan Meier, created a fictitious MySpace profile of a 16-year-old boy who seemed to want a friendship with Megan. After reeling her in, subsequent postings and emails turned hateful, psychologically devastating the girl, who had a history of depression and low self-esteem. On October 16 of the same year, she committed suicide. (http://abcnews.go.com/GMA/story?id=3882520&page=1)

Most cyberbullying, hurtful as it may be, does not result in death. Still, suicide is not an isolated result. There are many accounts with similar endings, and a great many more with damaging, though not fatal, outcomes.

The socialization debates   One school of thought is that people who make extensive use of social networks find themselves alone in the midst of a large circle of friends. In that view, the terms "social" and "friends" are seriously misleading: social requires in-person interaction; friend means much more than passing information back and forth electronically, no matter how intimately revealing that information might be.

Real-time audio/video links don't add much either. They are just a poor stand-in for face-to-face contact, for which there is no substitute. The social skills necessary for successfully navigating the complexities of life in the real world are not developed in virtual worlds. In the end, while technologically connected to many others, the person behind the interface is actually alone.

Opposing that view are those who hold that "social" and "friend" allow for many levels of engagement, from casual contact to deeply intimate. Some require physical presence, others do not. Isolation, on the other hand, means having no contact. Social networks are simply communication platforms. Quite the opposite of destroyers of socialization, they provide a significant means of increasing it. Rather than displacing in-person contact, they expand our range.

The result is that social skills are enhanced because there are more ways to socialize and thus to become adept at both geographically close and distant relationships. Encounters in cyberspace and face to face have important roles in our lives. A well-rounded socialized person is readily conversant in both.

Overlying these views are two related contentions. One is that social networks do not change our natures. If we are naturally gregarious, we will use social media boldly, not just to enlarge a circle of friends but to express our opinions and voice our beliefs. If we are reticent, we will be less apt to engage in those behaviors. The other is that people who would otherwise tend to be socially stunted when face-to-face is the only option find it easier to be more outwardly oriented by using social platforms. Anxieties associated with in-person contact are washed away online. That may even carry over to life offline.

Trust, an important element of most relationships, adds another dimension to the issue. We may be better able to judge trustworthiness in-person, using such clues as facial expressions, mannerisms, and vocal cues to gauge trust. These are missing from static online interaction. Even if real-time voice and video are incorporated, online judgment is not as reliable as face-to-face in-person contact.

Accurate judgment isn't guaranteed in either realm. In the hands of skillful manipulators, we can be beguiled into believing them quite readily in person and online. That's how pitchmen, con artists, and scammers make a living. The concern is that the more that social media become a routine part of our day, the more we may be likely to drop our guard.

All these contentions make valid points. We are an enormously varied lot, just as different in our use of social networks and social media as we are in our lives and just as different in our opinions. We are not fixed in one category or another for all time and in all instances and we have varying experiences throughout. We may find lasting friendships that exist only online, some that begin online and flourish in person, others that begin in person and continue online, and still others of every combination that fail. That's a mirror of life.

So we're back to "it depends." Social media are not the only way for us to express ourselves, but they certainly can be a powerful way. What we do with them is up to us. We may be true to our natures as we use social platforms or we may create alternate online personas. We may be out front or we may hide behind an interface. We may take advantage of the opportunities for beneficial socialization or we may use the platforms to perpetrate cruelties. Those too mirror life, amplified by digital IT.

What do you think about this? *Two views*

Violent video games, played alone or online with many others, dull us to violence and remove our inhibitions towards it. That carries over to the real world where, having become inured to violence, we are indifferent to violent acts by others and more likely to engage in violence ourselves. That's the opposite of socialization.

Violence in virtual worlds is an outlet for our violent tendencies. Playing those games diffuses our violent instincts and drains away our violent impulses. We become more empathetic to the victims of real-world violence and less likely to engage in violence ourselves. We are better socialized.

Business on board   Businesses are no strangers to social networking and social media. Exploring the complex picture that lies beneath that simple statement is beyond the scope of this book, but here are some important issues to consider.

Commercial and internal social platforms are being integrated into traditional business IT at every organizational level, from back office to frontline operations, from product and service advertising and sales to promoting the corporate image. An active online presence is part of almost every business model. Social platforms are a way to add to that presence.

Those developments have many pluses, but downsides are inevitable. Companies routinely scour social media for clues about the suitability of job applicants and the indiscretions of employees. Common off-the-job activities may be looked at as red flags. Histories, even if no longer relevant, can bite back. These contentious uses of social media easily cross over into invasion of privacy and violation of free speech rights.

As employees become more aware of business perusal of their social network pages, sentiment that their private lives are being confounded with their business lives is growing. Objections are raised that too much information about them is collected and held for too long, that they can't correct erroneous data or delete information they don't want corporations to have, and that they have little ability to prevent its collection in the first place. Furthermore, it isn't clear how carefully corporations protect the data from those who have no business seeing it and whether or not facts are checked.

On another front, employees who want to climb the corporate ladder have to be adept at using all the new media platforms and be readily available online as well.

Away on a business trip or vacation? That's no excuse for not being able to deal with a developing situation. Free time is harder to keep free; private lives are harder to keep separate and private.

Another issue derives from the growing inclination of employees to attend to their social networks while at work. That may mean using corporate facilities, thus adding to company network loads, or mobile devices connected to external services. Either way, attention is drawn away from work and productivity suffers.

As ever, there's another side to the story. If done at appropriate times and not to excess, the diversion of social networking can provide a much needed respite from the demands of work and actually make employees more productive.

Impacts rising

Social networks of one kind or another have always existed, but for the most part, their size, scope, and reach have been comparatively constrained. In the post-digital era, we've witnessed a tremendous growth in the number of online communities and the ease of finding, joining, and starting them. Accordingly, the number of participants overall is in the billions. Some networks themselves have many millions of members. That has greatly contributed to impact.

Socializing, in the sense of mingling, having fun, hanging out, and just idle chatting, is the main point of many social media. Facebook and Twitter are prime examples of those whose initial emphasis was socializing, acting as conduits for sharing information with self-created collections of contacts, friends, fans, and followers. With privacy set to allow wide access, collections become many-layered interconnected webs of relationships.

We've noted that on a technical level, Facebook, Twitter, and similar sites are merely applications that provide ways to interact and form differently focused networks of contacts and information sources. But we've come to see that they have significant value beyond the simply social. They have grown to such an extent that on every dimension—personal, business, political, and impersonal—their impact is very deep and complex.

A variety of purposes   Social networks can be effective instruments for carrying agendas forward. Just as physical communities do not need to be uniform in purpose, goals, or constituencies, neither do online communities. It's not necessary for everyone in a community to agree on a cause or an approach. On the other hand, as the number of participants who buy in increases, so does the community's influence.

Online communities can easily be formed on the fly and just as easily reformed and disbanded. Foundation goals can be the steady presence for which the community was formed. Ad hoc initiatives can meet challenges and needs important at the time.

For immigrants, social media offer support as general information sources, for connecting with people in similar circumstances and with common backgrounds, navigating the ins and outs of their new countries, finding work, dealing with bureaucracies, and more. They can serve as safety nets, providing practical advice, lessons from experience, sources of help, and guidance that otherwise might be hard to come by. They also offer fun and a relief from everyday problems.

Mainstream social media are often employed for those purposes. Users of Facebook and other general membership sites can create special interest groups within their structures that aim at particular constituencies, but that isn't the only option. Another is social networks designed from the start with a specific orientation to serve a particular community. Here are some diverse examples:

- *Dominican-diaspora.com*, a membership site, hosts forums, groups, chats, and blogs, has notices of events, provides relevant videos, and enables forming personal pages, among other features, all focused on Dominicans. As the site says, it's "A social network for people who live in the Commonwealth of Dominica or have family from Dominica. For those who live there, love there and have roots there!"
- *HaitianSpace.com* has features similar to Dominican-diaspora, in both English and French. It also has interactive blogs, currently in cultural, economic, political, and social categories. It was especially highly trafficked during the 2010 earthquake in Haiti.
- *MyChurch.org* offers three fee-based Facebook applications for churches: *Donate* enables congregants to make donations using PayPal and credit cards; *Prayers* allows users to request prayers and pray for others via the church's prayer wall; *Your Own Private Network* sets up a dedicated social network for the church and its congregants.
- *ASmallWorld.net* is, in their words, "the world's leading private online community that captures an existing international network of people who are connected by three degrees of separation . . . [the] unique platform offers powerful tools and user generated content to help members manage their private, social and business lives."
- *Sonico.com* is a Spanish language membership site that caters to all Spanish speakers but is targeted mostly to South Americans. It is one of the largest of the specific constituency sites, with over 8,000,000 members.
- *BlackPlanet.com* offers a variety of socializing opportunities, discussion forums, talent showcases, interest groups, and via members, links to other sites. BlackPlanet's "music, jobs, forums, chat, photos, dating personals and groups [are] all targeted to the specific interests of the black community."
- *Jmix.com* has similar services for Jews. "Mix, mingle and network with the Jewish community. Attend events and make new friends and business connections. Meet new people. Have fun!" the site proclaims.
- *Foodspotting.com*, a more mundane venue, goes beyond the typical restaurant recommendation site by drilling down to the dishes served, with photos and calorie counts accompanying customer reviews. The app with neighborhood searches is available for computers and mobile devices.
- *GlobalVoicesOnline.org* is a different sort of example, "an international community of bloggers who report on blogs and citizen media from around the world." Started by two Berkman Center for Internet and Society research

fellows in 2005,[10] the site is different from typical news scraper sites. Volunteers sort through content from serious bloggers posting on platforms in countries worldwide; all stories are subject to editorial review before being posted on GlobalVoices.

The founders' core values, reflected in the site, are based on belief in the importance of "free speech: in protecting the right to speak—and the right to listen . . . in universal access to the tools of speech." "To that end, we seek to enable everyone who wants to speak to have the means to speak—and everyone who wants to hear that speech, the means to listen to it." And so they look to create a global conversation forum.

Dozens of countries and languages are covered. Topics categorizing content run the gamut from agriculture to youth. Among the more active are cyber-activism, freedom of speech, governance, human rights, politics, and war.

● *SocialTattooProject.com* takes a peculiarly unique advocacy approach by aiming to make "empathy permanent. Volunteers agree to be tattooed with designs or hashtags that represent 'worldly issues.'" The twist is volunteers don't know what designs they're agreeing to have tattooed on their bodies. That's decided on by user response. "For each tattoo, we will post 4 trending topics on Twitter, and the most tweeted trend will be the subject of the tattoo." That's the so-called permanent empathy, whose expression remains even after the issue disappears. The site posts comments about the project, some supportive, many disparaging.

Politicians jump in   Politicos are no strangers to communicating with their constituencies, extolling presumably great accomplishments, seeking to turn voters to their side and convincing potential supporters to donate to their campaigns and action groups. Though not the first politician to utilize today's IT, President Barack Obama's extraordinarily effective use of email, the Web, and social networks during his campaign for the presidency certainly opened a lot of eyes to their potential.

Aside from being vehicles for disseminating his views and building the Obama brand, he was able to capitalize on social media's ability to help people organize their work in support of his candidacy. Importantly, all that was done at a much lower cost than would have been possible with traditional methods. He continues even more effectively on that trail, and now, so does First Lady Michelle Obama.

That did not skip by politicians unnoticed. They have seen the light and so have turned to mainstream social media and other IT in droves. These days, it's hard to find a politician at any level of government who doesn't have a presence on Facebook and Twitter to go along with a Web site, heavily used email lists, and even YouTube videos.

Impacts are not always positive, however. Increasing numbers of emails from political operatives, action committees, and the like are reaching a point where we routinely ignore them, just as we are numbed to the postal flood of campaign literature and blight of roadside posters that accompany elections. Firing up supporters and attracting others to the cause requires a delicate balance in drawing a line between the information quality and quantity that excites and that which causes withdrawal.

The more that political operatives contribute to the flood, the more frequently the line is crossed and the less effective the pleas and boasts.

Predictable? Perhaps   We've come to see the impossibility of keeping secret anything that's online in one form or another. These days those forms often are social media sites, blogs, and records of cell phone texts. Yet despite mounting evidence, people and governments alike continue to experience impacts of posts, indiscretions, and even the most innocuous of events. Consider these.

- An August 2011 photo of newly appointed US ambassador to China Gary Locke and his daughter at the Seattle-Tacoma International Airport went viral after it was posted on the popular Chinese social network Sina Weibo by the Chinese-American who snapped it. The mundane photo showed Locke wearing a backpack and buying coffee at the airport Starbucks. To mainland Chinese, used to hearing about the lavish perks and privileges of party officials, the appearance of an American ambassador in an ordinary scene was striking.
- Adding to the impact was another photo of Locke and his family taken on their arrival at Beijing Capital International Airport that showed them carrying their own luggage. Chen Weihu, an editor at *China Daily* said, "To most Chinese people, the scene was so unusual it almost defied belief."[11]

These photos had greater impact on the Chinese than formally arranged events, especially since the citizenry has been increasingly upset with party officials who they see as out of touch with everyday living in China and whose lifestyle is far beyond the reach of most of the population. Thanks to the speed with which information travels on the Internet, the photos spread before any attempt at containment could be mounted. Government members have to take note.

- Also in China, the tradition of athletes' quiet obeisance to national sports authorities was rattled in April 2011 when members of the Chinese national junior basketball team sent a letter to the Chinese Basketball Association outlining grievances against their coach. The letter, leaked to the media, resulted in the coach being disciplined and became an impetus for young athletes to speak up for themselves. "[Now] the younger generation of athletes has so many options to communicate, through microblogs and social networking, that they want to stand up and speak out."[12]

Germany offers another case in point.

- Forty-year-old state legislator Christian von Boetticher became one more of a panoply of politicians caught by social media. He had a fairly successful political career despite a history of Facebook postings that many of his colleagues felt were far too extensive and revealing. The proverbial straw that broke the camel's back was the coming to light of his affair with a 16-year-old girl he met on Facebook. Though 16 is the legal age of consent in Germany and though

Germans tend to separate the sex lives of politicians from their official conduct, in sum there was too much heat. Boetticher resigned in August 2011.

Boetticher's indiscretion, just one in a long line of similar activities, learned that sooner or later, evidence and details out on the Internet come back to haunt. It's quite likely he won't be the last one caught in the same way.

Specially for you   Several Web sites have sprung up offering applications for people who want to create their own social network sites. They have the features of true social media, quite different from personal Web pages or blogs.

Ning.com, rSitez.com, and SocialGo.com are three examples. Each offers easy-to-use do-it-yourself application tools to build a social networking site. Each has a fee structure based on site complexity, hosting, and data storage. If you'd rather have them build your site, they do that too, for a fee.

As these and other such sites become more capable, small businesses and corporations are turning to them. Costs are low and they require little in the way of in-house maintenance. So businesses find them especially useful for ad hoc campaigns, trying out new product ideas, and building public interest in various outreach programs.

Upending the social order—a spark and sometimes a flame

Upending the social order—a spark and sometimes a flame   We frequently note that technology in general and IT in particular have unintended consequences —uses and results not envisioned when the technologies were planned or introduced. Recent uprisings in the Middle East, the so-called Arab Spring, are an excellent example, having done for social protesters what Obama did for politicians in utilizing IT.

Thanks to autocratic control over mainstream media—radio, television, newspapers, movies—and strict prohibitions against criticism of leadership, the populace in countries with oppressive regimes have long lacked the means to coordinate effectively to overcome barriers to voicing their opposition and exposing their true plights to the rest of the world. The Internet and wireless communication changed that. Now it is increasingly likely that breaking news comes from people in the middle of the action before professional reporters are on the scene, especially when the location is announced with very little advance notice.

We look at the impacts of social media and wireless communications with brief summaries of conflicts in several countries. These overviews do not fully explain what happened and why, nor are they meant to. Furthermore, the statuses of countries noted are evolving daily and as of this writing have yet to reach substantial stability, so situations are likely to have changed. However, the examples serve to illustrate the roles played by IT in the uprisings, successfully and not. They also reveal that government hegemony and stability can be threatened by newer communications technologies in the hands of only a very few people. That's a new development.

Tunisia was the prime mover in a wave of uprisings that swept the Middle East. A litany of abuses—corruption, severely restricted freedoms, subpar living conditions, high unemployment, and rapid inflation—set the stage. In December 2010, demonstrations started a groundswell of continuing protests that turned civil unrest into civil resistance. On January 14, 2011, after about a month of escalating protests, the government of President Zine El Abidine Ben Ali capitulated, ending his regime of over 24 years.

The initial demonstration was sparked by videos and reports of the self-immolation of Mohamed Bouazizi, a 26-year-old street vendor abused by the police whose plight was ignored by municipal officials. News spread quickly via Twitter, Facebook, and YouTube. At the same time, WikiLeaks published reports of government corruption. Using the Internet and cell phones, protesters were able to organize quickly, keeping ahead of major police actions. The extent of the protests grew rapidly even as demonstrators were shot and jailed.

The actions of the Tunisian people showed the world what civil resistance could accomplish in countries where public displays of disobedience had been successfully quashed for many years. A key difference was the effectiveness of IT for coordinating the demonstrations and disseminating descriptions and videos of the harsh and lethal attempts by the government to stop them. The success of the uprising was not lost on the citizens of neighboring countries. Informed of what was happening in Tunisia by social media, cell phones, Al Jazeera,[13] and other IT platforms, they followed the example.

Next up was Egypt. Barely two weeks after the downfall of the Ben Ali regime, Egyptians, whose grievances were similar to those of the Tunisians, organized quickly via Facebook and Twitter. Many thousands filled Tahrir Square in the center of downtown Cairo, the capital.

Photos of the Square, posted on Reddit[14] the day after the first demonstration, went viral, That helped the rebellion by bringing the attention of the world into play. As the size of the protest grew, the government responded with violent police actions and arrests of hundreds.

In an attempt to disable the ability of protesters to organize, the government shut down Egypt's main cell phone carrier, Mobinil, blocked Twitter, and suspended ISP access accounts. Workarounds and technical help from people in other countries kept communications flowing enough for organized protests to continue and for news of the uprising to reach outside Egypt. Then too, demonstrations fed on themselves, drawing more and more of the populace to the cause.

Demonstrations, work stoppages, and other disruptive tactics soon spread to Alexandria, Egypt's second largest city, and then to others. Eventually, protesters numbered in the millions. On February 11, only a little more than a month after the demonstrations began, President Hosni Mubarak resigned, ending his 30-year reign.

An unexpected use

SayNow, a successful startup purchased by Google, began as a means for celebrities and sports figures to leave short voicemail messages for their fans. Google used it to provide a service whereby Egyptians could dial international phone numbers and leave voicemail that was converted to #egypt hashtag tweets in Google's Speak2Tweet Twitter account. So the latest news of the uprising, including government reactions, still spread to the world after Internet access was shut down.

What do you think of this? *Why now?*

Revolutions take place in the streets, not in the Internet. So why did these uprisings not happen earlier in the 24 years of Tunisia's oppression or the 30 years of Egypt's?

The turning point was the ability of the populace to organize quickly and effectively. That was the role played by Facebook and Twitter, cell phones and satellite phones. Of course, a full explanation is not that simple. The point is that though social media and wireless communications did not cause the revolutions, nor the overthrow of the governments, they gave compatriots the means to mobilize the citizenry, focus anti-regime sentiment, organize to take their grievances to the streets, and show the world what was happening. Those are powerful abilities.

Lesson learned?   Citizens of other autocratic countries watched events unfold in Tunisia and Egypt, and so did their leaders. Both sides saw the effectiveness of IT platforms in uniting opposition and organizing protests to spark movements that took off. It was clear that use and control of communications mechanisms were an integral part of the battles. One thing apparent to all—once demonstrations gained sufficient momentum to reach a critical mass, they were very difficult to contain and even harder to stop. So sovereigns have taken two courses of action to retain power—appeasement and crackdown.

Some rulers attempted to short-circuit protests by offering concessions aimed at damping down enthusiasm for demonstrations. After all, despite the Tunisian and Egyptian outcomes, there is no guarantee that success will follow from an uprising. Moreover, protests are stressful and physically demanding. Participants have been jailed, injured, and killed. There also is the possibility that failure can bring with it an even more restrictive regime.

Threatened with loss of control, some governments have reacted with deadly force to cripple protest movements before they could gain traction and develop into mass demonstrations. Widespread condemnation and pressure from many of the world's countries has not made much difference. Here are some major recent examples.

In Yemen, social media played the same role as in Tunisia and Egypt—for galvanizing and organizing protests, and publicizing pictures of demonstrations, government crackdowns, and funerals of protesters. The results, though, were quite different.

About the same time as the Egyptian uprising took off, over 16,000 demonstrators took to the streets of Sana, Yemen's capital city, calling for President Ali Abdullah Saleh to resign. In an effort to mollify the populace, Saleh stated he would not run for reelection in 2013, nor hand the presidency to his son. That was seen as too little too late, especially since elections were considered controlled. Protests continued to grow, spreading to other parts of the country. By March, violence entered the picture when pro-government forces killed more than 50 protesters.

Negotiations with the government went nowhere as violence escalated. Both sides used artillery and mortar fire, culminating in opposition forces bombing the presidential compound, killing several people and injuring Saleh. Still, he refused to back down even after many of his government officials resigned in protest of the violence he precipitated.

Despite opposition gains and army defections, by dint of his intransigence and continued control of significant components of Yemen's armed forces, Saleh had been able to stave off his overthrow. Even though he signed an agreement in late November whereby he would leave office by the end of the year, it was not accepted by all sides in the conflict. So we see that as powerful a tool as IT is, it does not guarantee anything.

Bahrain, by far the smallest Gulf kingdom, has a population of only about 1.26 million, of whom about a quarter are immigrant workers. It gained independence from Britain in 1971 during the reign of Isa bin Salman Al Khalifa, but was ruled by the Khalifa family much before that and still is. Hamad bin Isa Al Khalifa, son of Salman, has been king since 2002. Khalifa family, inheritance, and relationships dominate many government posts.

Among other societal issues, religious conflict is a chief source of discord in Bahrain. The Khalifas are Sunni Muslims, but about three-quarters of the Muslim population are Shi'as. They are not happy with Sunni leadership, which they perceive as being discriminatory towards them.

Rivalry between religious factions has been a major driver of much of the conflict in the Middle East. Iran and Saudi Arabia are prime examples—over 99 percent of Iran's population is Shi'a while Saudi Arabia's is about 85 percent Sunni. The two countries are carefully watching the conflict in tiny Bahrain as a possible preview of things to come. This is especially important to Saudi Arabia since it and Bahrain are monarchies.[15] So it is not only a question of religion, it also is one of power.

The Bahrainian uprising followed a pattern similar to Yemen's. Seizing on social media as an organizing and publicity tool, demonstrators began protesting in February around the same time as in Yemen. It also turned violent, with a severe crackdown by the police, many arrests, torture, and deaths. It has not resulted in the reforms that were sought, which initially included equal rights and more political freedom of choice, later escalating to a demand for relief from strict Khalifa control.[16]

Once again we see that as news of successful protests spread, others used IT to organize their own movements, and once again we see the ability of quickly employed superior physical force to quell demonstrations and maintain the status quo.

Libya gives us another perspective, though. Also in February, the citizenry began an effort to reorganize Libyan society. The same progression of peaceful demonstrations turning violent ensued, but to a much greater degree. The result was an out and out civil war against the rule of Muammar Gaddafi,[17] in power since 1969 when as a 28 year old he led a group of military officers and soldiers in a bloodless coup that ousted King Sayyid Muhammad Idris.

Weapons used by both sides ranged from pistols to heavy artillery, tanks, land mines, explosives and other assorted armaments. A point of departure was that many of the world's countries, galvanized by online and offline media reports of the crackdown, supported the rebellion. A United Nations resolution authorized member countries to create a no-fly zone over Libya and to bomb certain vital military installations to hamper Gaddafi's ability to attack the rebels. Eventually the rebels gained the upper hand, having recaptured the oil fields and the capital, Tripoli.

Gaddafi disappeared early in September, but he didn't leave the country. Late in October he was found and killed. Despite the success of the uprising, disputes among factions within the rebel groups leave the ultimate outcome unclear.

IT played a role for both sides. The rebels took the familiar path of organizing proponents and publicizing victories as well as the destruction and deaths caused by Gaddafi's response. Libya's official press, radio, and TV media controlled by Gaddafi took aim as well. Rebels were portrayed as rats, agents of outside forces, and conspirators who didn't care for the people of Libya but were only bent on its destruction.

At the same time, blockages of civilian communications and Internet access made it difficult for opposition forces to coordinate their actions and spread word of government violence. Incoming news from Al Jazeera and other broadcasters was hampered by destruction of the satellite dishes needed to receive their signals. Systems restored in areas controlled by the rebels offset these moves somewhat.

In Syria we see yet another take in a country long overseen by a single family, the Assads. Bashar al-Assad, president since 2000, followed the presidency of his father, Hafez al-Assad, whose reign began in 1970. The progression of the uprising somewhat parallels Libya's.

It began quite modestly after a self-immolation on January 26, 2011, similar to what happened in Tunisia. Outraged Syrians took to Facebook and Twitter to spark an uprising. They called for "Day of Rage" rallies on February 5 and 6 to demand government reform but stopped short of demanding Assad's departure. The Day attracted more attention outside Syria than within, where only a few hundred attended. Security forces quickly stepped in, ending the rally and arresting several dozen demonstrators.

Discontent simmered below the surface for a month and a half, when Facebook and Twitter again were used, this time more successfully. Thousands of demonstrators in several cities participated on March 15. On March 25, more than 100,000 people rallied in Daraa, a city whose population is only about 77,000. Syria's leaders faced a full-fledged uprising.

The government reacted by an escalating series of crackdowns. Many people were killed and many more were arrested. As demonstrations spread, government violence increased. Syrian army members on the ground, accompanied by helicopter gunships, fired on crowds of protesters. Many soldiers who would not shoot civilians defected from the army; others were executed.

The government also counterattacked via the media. Dismissing reports of defections, they blamed the violence on gangs of armed thugs that the government had to put down to preserve the peace. Photos and videos of large-scale demonstrations in support of the president were posted to bolster their case that the revolt was not a popular one.

Violence by government forces continued to escalate. Blockades isolated cities, stopping incoming supplies. Power and communications were cut. In Hama, Syria's fourth largest city, only about 120 miles north of Damascus, the capital, a prison rebellion added to the turmoil of demonstrations. Both were met by government-ordered artillery and rifle fire.

Faced with growing international pressure on the government to end attacks on civilians, state news media sought to paint a different picture. Photos showing a quiet Hama after the uprisings were quelled were displayed with accompanying reports as evidence that the rebellion had ended. Once again the upheaval was claimed to be the result of armed gangs and saboteurs under the sway of foreign powers with their own agenda, who did not have Syria's interests at heart. The government claimed that by restoring order, the army was the savior of Syria and its people.

In Iran, stirrings of unrest and attempts at demonstrations were quickly met by strong force to suppress any opposition before it had a chance to root. Hundreds of protesters have been executed. Tight control over information flow, including shutdowns of newspapers, arrests of journalists, and monitoring of blogs and texts to find dissidents, added to the picture. Sophisticated Web site filters blocked sites considered objectionable. Meanwhile, the state-controlled newspaper *Hamsharhri*, along with state radio and television spread the government's messages.

People outside Iran lent their voices after Neda Agha-Soltan, not yet part of a protest of Iran's 2009 elections, was killed by a government militiaman. Cell phone photos and videos of her dying in the street rallied protesters in Iran and around the world. That sparked Iranian expatriates to take advantage of Internet platforms to express opposition to government control of media, harsh anti-demonstration tactics, and lack of freedom of speech. Some pushed for significant political reform, open elections, and reduced religious influence over political leaders and reformists.

That has not had much effect inside the country for now, in large part because of the government's success in blocking access to information combined with rapidly applied force to stifle demonstrations. Still, it's another IT tug of war that may yet shift away from government control.

Not only in autocratic regimes   Israel, the only fully democratic country in the Middle East, is one where freedom of speech and open elections are rigorously endorsed. There are few instances of the ills that fostered anti-government demonstrations in the region. Yet Israelis, growing increasingly upset by the country's lopsided distribution of wealth[18] and a cost of living that is progressively more out of line with incomes, have followed the protest model and taken to the streets.

With unfettered access to the Internet, the media, and other communications systems, protesters easily organized demonstrations in several cities in August 2011. Hundreds of thousands participated in repeated demonstrations that spread across the country. Prime Minister Benjamin Netanyahu pushed for talks but did not take any steps to cut off communications or otherwise stifle the demonstrations, which have remained peaceful.

In England, another democratic open society, the story was different. Demonstrations began in March 2011 to protest the government's planned cuts in public spending that many believed fell most harshly and unfairly on union members and those in the lower income strata of the country.[19] There was no clear count of the number of participants, though all estimates are in the several hundred thousand range.

As was the case in Israel, open access to communications systems made it easy to organize the first rally and those that followed. But peace did not reign. In August, police in north London, attempting to disband a peaceful demonstration, were set on by some of the protestors. Presumably that turn to violence was the culmination of ill will towards the police that had been swelling since the fatal shooting of a local resident two days earlier.

Subsequently, violence spread across the country as a great number of people, mostly youths, took the opportunity to damage cars and buildings, loot stores, and set fires. It's not clear why that happened, since the shooting was an isolated act and the riots had no obvious link to the initial rallies. Aided by ready access to Twitter, Facebook, and cell phones, the speed with which violence spread caught the police off guard. In the chaos that ensued, people on both sides were injured. Numerous looters were arrested as were some bystanders who were caught unawares.

The destructive acts turned many against the rallies. Rioting spread faster than the original protest movement, which lost steam due to the overwhelming impact of escalating destruction. Twitter became indispensable to frightened residents, who used it to keep up with the outbreaks so as to stay a step ahead, out of harm's way.

Hoping to gain a political advantage, leaders of opposition parties took the opportunity to blame Prime Minister David Cameron and the Conservative/Liberal Democrat coalition for the condition of the economy, the demonstrations, and the riots. Under pressure from Cameron and the chaos in the streets, police took to monitoring Twitter and BlackBerry messages to get information about the next riot sites. That was a significant step in a country that places a high value on privacy and information confidentiality. Even more dramatic and disturbing was Cameron's call for social networking sites to block access to those who use them for encouraging and coordinating acts of violence. That can't be done without scrutinizing posts and tweets.

In the US, a different sort of protest movement arose, intensely supported by IT— Occupy Wall Street (OWS). Two events were precursors to the actual movement. One was a July 13, 2011, blog post by the Adbusters Foundation[20] advocating a peaceful demonstration on Wall Street to protest undue corporate influence on the government. The other was the call by Anonymous [21] to its followers to join in a protest, occupying Wall Street by setting up encampments in the area.

On September 17, 2011, the actual movement began with a large peaceful demonstration in New York City's Zucotti Park[22] just north of Wall Street. Their slogan, "We Are the 99%" pointed to the growing disparity in the distribution of wealth in the US, with most in the hands of just 1 percent of the population. They also railed against the disproportionate influence on government of giant corporations and the financial sector, high unemployment, corporate greed and corruption.

Several groups joined unaligned individuals as the movement grew. Within days the occupiers had set up a Facebook page. They and many others not physically involved also posted YouTube videos and an increasing volume of tweets with the #OWS hashtag in support of the protestors. More or less leaderless, they set up the General Assembly as an organizational body, though it often relied on protestor consensus and the work of a number of volunteers who took care of routine matters like food procurement and distribution, finance, sanitation, and interaction with the police and the media.

The protestors occupied the park 24/7 until mid-November, when by request of Brookfield based on claimed hazardous and unsanitary conditions in the park, the police ordered the demonstrators to leave. Most did and the few that didn't were arrested. They were allowed to return after the cleanup, but could no longer stay overnight.

The direct influence of the movement on Wall Street firms is hard to gauge, though apparently no more than minimal at this point. But the bigger picture shows much greater impact. The movement spread in two ways: similarly styled protests focused on a variety of issues; worldwide expansion. In the US, OWS spread cross country to dozens of cities. Worldwide movements have agendas related to and different from those of the US OWS, but with a similar organizing and spirited means of protest. A dedicated Web site (http://occupywallst.org/) tracks all related national and international developments.

And so—the many directions of impact

> New communications patterns have now been woven into group life. The activity at organizations and informal groups is increasingly taking place in social networking spaces like Facebook, on Twitter, in blogging, and in cell phone texting.[23]

> Quit counting fans, followers and blog subscribers like bottle caps. Think, instead, about what you're hoping to achieve with and through the community that actually cares about what you're doing.[24]

Social media and other forms of digital communication have become integrated into societies worldwide. Social networking has gained status as an important means for broadening horizons in ways and to degrees not otherwise realizable, including tuning in on opinions and developments worldwide.

The flip side of being vastly more connected than we've ever been is that in the process, we've traded some amount of in-person contact for a much greater amount of online contact. We've also exchanged privacy for ease. That has inured many of us to the intrusive collection and use of personal data practiced by businesses, organizations, governments, and the very sites that give us those wonderful connection opportunities. And so, we pay less attention to those practices.

Another issue is extension of the digital divide. Formerly it primarily focused on access to computers. Now, access to mobile communications has become at least as important, if not more so. In the age of social media, that is especially significant for underdeveloped and developing countries, where physical travel can be difficult if not impossible. Lack of access obviates opportunities for residents on the wrong side of the divide to become integrated with societies beyond their local communities. Mobile and other wireless systems can be simpler and less costly to establish than cable-based systems, so they can be the best way to reduce the mobile divide.

Moving up the scale to cooperative participatory movements, we find that social media are increasingly used as platforms for advocacy. They enable activists to create forceful coalitions to gain political influence, to change the direction of governments, and to begin movements to overthrow them. Effective in spreading information and organizing efforts, demonstrators in closed and open societies alike use them to good advantage.

Of course, these platforms are not just the province of activists. The same applications give autocrats and freely elected officials organizational and information distribution abilities as well. National leaders and other government officials also form coalitions, rally supporters, and spread their views on the state of their nations, public actions, and the future of their countries. Battles fought in the new digital media are apt and powerful parallels to battles fought in the streets.

Implications for control No matter where it's done, even in the most tightly controlled countries of the world, the monitoring of messages, calls, postings, tweets, and the like has to be selective. Even with computerized sniffing algorithms and legions of volunteers, everything from everyone can't be checked all the time. So agencies are forced to focus on known activists, dissidents, and those they rightly or wrongly consider to be suspects.

That means missing all the newcomers to the game and probable intrusion on the lives of some who are not a threat of any kind. Then too, activists can take measures to disguise their online identities and transmissions, making it that much more difficult for agencies to get relevant information in time. That's a plus in oppressive societies and a bane for open ones whose efforts are aimed at protecting the populace from true dangers.

Aside from seeking information about destabilizing or destructive events beforehand, surveillance also can lead to access cutoffs. Juxtaposed to shutdowns by rulers in autocratic regimes who are excoriated for those actions by leaders of open societies, the calls in England for surveillance and cutoffs—even though comparatively slight and not enacted—are astonishing. That highlights an important issue for open societies—Are there times when curtailment of civil liberties is justified, the proper course to take?

What do you think of this? *Prevention, intervention, or intrusion?*

Tackling quality of life issues, the New York City Police Department established a unit to "track troublemakers" by searching social media for clues to gang activities, boasts of criminal mischief and worse, ordinance violating parties and gatherings, and other socially disruptive and illegal activities. The idea is to stop behaviors ranging from bothersome to criminal before they happen and to find perpetrators after the fact.

Source: http://www.nyc.gov/html/nypd/html/crime_
prevention/crime_prevention_section.shtml

The value of putting the power of social media to use in aid of law enforcement goes without saying, as does the value of preserving quality of life. But what about privacy and freedom of speech? Drawing the line can be a delicate balancing act. How do you feel about any policing authority following and in some way discouraging people who might engage in annoying or even criminal activity before they actually do?

From the bottom—replacing formal organizations Demonstrations have a long history of being used effectively by activists of all stripes. The contribution of the digital era is the speed with which movements can come together and that is due to the many digital communications platforms available.

One reason that demonstrations organized via Facebook, Twitter, and cell phones have been so effective is that they typically are not the product of predefined agendas and rigid organizational structures, whose actions take a long time to develop. Instead, information flows quickly, attracting people who share grievances even if they have nothing else in common. Simmering feelings of injustice can be readily tapped to form ad hoc groups informally and on short notice. Protest marches can be called on the fly.

Lack of meticulous preplanning adds an element of unpredictability, making it harder to control what seem to be spontaneous outbreaks, much like flash mobs. Key for these movements is gathering enough momentum to become self-sustaining, with growth more difficult to contain. That's the people part, the feet on the ground, the protesters in the squares and parks. Once that happens, flash is less relevant.

What do you think of this? *In a flash*

Flash mobs began about a decade ago to call people to participate in spontaneously formed events that could be as innocuous as a mass dance or sing-along. Soon after a mob meets, it quickly fades away. Usually formed via cell phones and social media posts, flash mobs appear without warning, seeming to come out of nowhere, and vanishing just as mysteriously. That, in large part, is the very point of the flash.

The fun side of flash mobs came next, as they morphed into calls for large assemblages to perform odd and meaningless acts. In their latest incarnation, violence entered the picture. Rampaging flash mobs have popped up in major cities across the US and other countries, looting, committing assaults, property destruction, and general mayhem.

The issues: Should social media and cell phone networks be monitored to alert police of impending mob formation or should communications privacy be upheld? If the former, how should the people and the platforms to monitor be selected? If the latter, does that mean preemption is impossible, leaving security and protection compromised?

Spreading the news Cell phones with cameras are commonplace, putting technology that used to require bulky equipment and a professional crew into the hands of ordinary individuals. These phones have been used to document protest movements and government reactions, the plight of citizens in countries torn by tribal warfare, famine, and disease, and the efforts of relief organizations, successful and stymied. That is a real-time form of testimony.

Photos and short videos posted on social networks and other media sharing sites are a quick and easy way to get the word out. Anything dramatic also gets picked up by aggregator sites and traditional media, further spreading the news. By showing the world what's going on, one person with a fairly simple phone can greatly influence the course of events. Imagine what can happen when hundreds or thousands of people do the same. Containing the news and squelching testimony to events is next to impossible.

Photos and videos of the 2005 London subway bombings taken by people who were in the stations and trains at the time provided first-hand views of the destruction and chaos that otherwise would not have been captured. The images helped the push for greater security and anti-terrorist activity.

Cell phone cameras recorded dramatic scenes of the effects of the 2010 tsunami that devastated Sumatra, causing widespread damage and deaths. Videos and photos taken on site during and immediately after the tsunami had a huge impact, even greater than the shocking views of the aftereffects. The same can be said of the 2011 tsunami that hit Japan. Posts on hundreds of sites were picked up and shown innumerable times by traditional and online media, galvanizing people and governments around the globe.

All is not positive, however. As we've seen over and over, people can use technologies in many ways. Phony photos and videos made to look authentic can spread just as easily as genuine ones. Authentication requires some due diligence. Before news media and aggregator sites post any material, some time should be spent seeking corroborating information. That means overcoming the urge to be first and fastest, which can result in spreading false information. On the other hand, the impact of immediate reportage is lost when information is delayed, yet another tradeoff issue.

Too much of a good thing?—balance, please   We've moved to a time when keeping in touch is as simple as sending an email, tweeting, and updating a Facebook post, among many other options. Whether we're aiming at one person, hundreds, or multiple thousands, it's delivered at the same time with hardly any delay, globally, at very little cost. Mobile technologies make it even easier, freeing us from having to be at a specific location to send or receive information, for direct contact with someone, and for connecting to the Internet and the Web.

These are wonderful abilities, but have they changed us? Many of us feel pressure to amass hundreds of friends and more. We've created a mindset of having to be constantly connected, continuously updated, never out of touch. We can't go long without checking our messages. We must tell our followers what we are doing at any particular moment, no matter how inconsequential, and those whom we follow must tell us.

Information sharing, intentional and otherwise, has become the norm—what's mine is yours and vice versa—boom times for social engagement. Privacy has become secondary, an inconsequential concern for many of us. We routinely broadcast our locations and we're tracked even when we don't. We have hundreds or thousands of friends, most of whom we don't really know. We are socially surrounded and involved in virtual spaces while we are physically alone.

There's no denying that social media are an excellent way to keep in touch, to rediscover former friends and acquaintances, to make new connections, to strengthen family circles, and to have fun. We do that with a growing number of platforms, but how many do we need? The more we use the more we must keep track of and update. How many tweeters and Facebook fan pages can we really follow? How many groups can we actually participate in? How many foursquare mayorships can we pursue? How much information can we contribute?

Many of us are facing social network overload. We've compounded the notion of information overload, largely concerned with how much we can absorb, with participation overload, which adds how much we can contribute. Put our ever present mobile devices into the mix and we also must answer calls, check and send emails, get the latest stock market quotes and sports scores, and jump into our social platforms on the go. At some point we are saturated. Beyond that, we fight to keep from drowning.

In other chapters we've seen that productivity falls when we have too many choices to make and attention falters when we face too much detail. The parallel here is being on too many social media sites, participating in too many feeds, feeling pressure to always be connected. Applications that seek to rescue us by presenting sweeps of our site flows and distributing our contributions over several sites are not much help. The trick is to strike a balance.

Relief can be as simple as focusing our energies on one or two social media sites and paring down groups to those with which we really want to be in routine contact. It's another question of where to draw the line—this time between fulfilling participation and its

> ## A little forethought goes a long way
>
> When email first began to catch on, it was so easy to dash off something and click send that we often didn't take time to read it over. Too late we've had second thoughts about what we wrote. Once something is sent, it can't be taken back. So we've learned to be a little more careful, reading over what we write before we click send, at least most of the time.
>
> Users new to Twitter, Facebook, and the like, also have felt the shock of sending or posting first and regretting later. Experienced users too can be careless at times. The essence of tweets is short and fast. Reading before tweeting just slows us down. That makes us more likely to click before checking than when writing emails. The lesson is the same—resist the urge to click until you're sure it's what you want to say.

diminishment from too much too often. When being connected becomes an obsession, when eagerly seeking every update, when being without a link is like withdrawal from an addiction, the line has been crossed. Remember, every device has an off switch.

The ball is in our court. We've seen how social media and mobile communications have had impacts on levels ranging from simple fun to deadly serious revolts, from expanding our associations to isolating us behind our devices. Most importantly, we've come to realize that social networking is not just about technology, it is mostly about us. We choose to use those IT tools as reflections of ourselves and as curtains to hide behind, to develop our true social circles and to surround ourselves with minions of casual contacts, to play games and to organize demonstrations, for social good and social evil. The choice is ours.

Notes

1   Jordan Raynor, *Online Strategy Fuels Brown Surge,* January 19, 2010: http://www.jordanraynor.com/2010/01/

2   Denise Zimmerman, *7 Things Social Media Can't Do,* February 22, 2010: http://www.imediaconnection.com/25979.asp

3   http://www.bbsdays.com/people/ward_christensen/

4   To transmit, a modem (**mo**dulator/**dem**odulator) converts digital computer output into analog data and sends them over a telephone line; a modem on the receiving end converts the analog data back into digital data for input to a computer.

5   ERIC Identifier: ED278381, Publication Date: 1986–07–00, Author: Manning, Ric, Source: ERIC Clearinghouse on Information Resources Syracuse NY: http://www.ericdigests.org/pre-925/computer.htm

6   Fidonet, having evolved considerably, is still operating. See: http://www.fidonet.org/

7   To travel over the Internet, Fidonet used UUCP. See *The UUCP Project:* http://www.uucp.org/

8   AOL (America Online) began in 1989 as an ISP and provider of online services. At its peak it had over 30 million members worldwide, but an ill-fated merger with Time Warner in 2001 presaged its decline.

9   ICQ was an instant messaging service created and marketed by Mirabilis.

10 Rebecca MacKinnon, former CNN Beijing and Tokyo Bureau Chief, and Ethan Zuckerman, technologist and Africa expert. For more about Rebecca MacKinnon, see: http://rconversation.blogs.com/about.html. For more about Ethan Zuckerman, see: http://ethanzuckerman.com/ For more about the Berkman Center, see: http://cyber.law.harvard.edu/

11 Quoted in "Photo Turns US Envoy into a Lesson for Chinese," *New York Times,* August 18, 2011, p. A4.

12 Quoted in "Saying No to the System," *New York Times,* August 19, 2011, p. B10.

13 Al Jazeera, based in Qatar, is a broadcast arm of the Qatar Media Corporation, which is owned by Qatar. Yet Al Jazeera is seen as independent and noted for straightforward broadcasts of current events. Its programs are sent over TV, satellite, and the Internet.

14 Reddit (reddit.com) is a social media site owned by Condé Nast Digital, which itself is owned by Advance Magazine Publishers Inc. It publishes content and comment from submitters worldwide.

15 Saudi Arabia is linked to Bahrain by a causeway that was constructed in 1986 to provide a quick route to Bahrain in case the Khalifas needed help to stay in power.

16 Some advocated diminishing the power of the monarchy by having it cede considerable authority to a freely elected parliament; others wanted an end to Khalifa rule, especially King Hamad as well as Prime Minister Khalifa ibn Salman Al Khalifa, who has been in power since 1971.

17 There are several versions of the spelling of his name, including Gadhafi, Gaddafi, Gathafi, Khadafy, Kadhafi, Kadafi, and Qaddafi.

18 The gap between the rich and poor in Israel is one of the largest of any of the 34 members of the Organization for Economic Cooperation and Development (OECD), whose mission is "is to promote policies that will improve the economic and social well-being of people around the world" (http://www.oecd.org).

19 Planned cuts in public spending amounted to 80 billion pounds (about $130 billion) by 2015, aimed at reducing huge deficits resulting from bank bailouts. Large cuts in welfare and elimination of thousands of public sector jobs were particular points of protest.

20 A Canadian anti-consumerist organization and publisher of an ad-free magazine espousing its causes.

21 Anonymous is a widely dispersed loosely organized activist group that utilizes the Internet to undertake and incite various forms of civil disobedience, including hacking.

22 Zucotti Park was Liberty Plaza until 2006, when Brookfield Properties took ownership, rebuilt it for public access but retained private ownership, and named it after its chairman, John Zuccotti.

23 Lee Rainie, Kristen Purcell, and Aaron Smith, Pew Internet and American Life Project report, *The Social Side of the Internet,* Section 3, January 18, 2011.

24 Amber Nasland, "10 Ways to Get Serious About Social Media," *Social Media Today,* January 3, 2010: http://socialmediatoday.com/index.php?q=SMC/161834 OR brasstackthinking.com

7

Attacks, bit by bit—from nuisance to cyber warfare

If you reveal your secrets to the wind, you should not blame the wind for revealing them to the trees.[1]

When did the future switch from being a promise to being a threat?[2]

The digital world is made up of bits represented by pulses and patterns of electricity, radio waves, microwaves, and infrared light. Bits have no intent, no morality, no goals. It is what we do with them that makes all the difference.

Zooming along the Internet and other digital pathways, packets of bits may be distinguished by packaging indicators—this packet has priority but that one can be delayed, the next follows a predetermined path while the one after needs its next hop calculated. But in terms of packet content and purpose, bits are just undifferentiated traffic. So it doesn't matter to the Internet if a packet is part of a transmission that carries an image of a chicken casserole or a pornographic photo, disk cleanup software or a virus.

This is analogous to vehicular traffic on roadways. That emergency vehicle has priority but the one behind it can be delayed. The driver of this vehicle is following a pre-mapped route while the driver of the other one is figuring out the next turn on the fly. The truck ahead may be carrying food or bombs. Passengers in the car alongside may be on the way to a party or a robbery.

Within the Internet, traffic flow volume is monitored to try to keep pathways uncongested, but it is only at particular points—mostly Internet entryways and exits—that content and intent can be determined. That's where businesses erect firewalls to thwart incursions and monitor incoming traffic to stop Web site-crashing overloads, where equipment has antivirus software installed to prevent infections, and where Internet service providers (ISPs) attempt to filter out spam.

There are a lot of people who want to make the world a better place and a lot who want to make it worse. Computer-chip based devices and an Internet connection form a major doorway for both sorts of actors. That's a simple idea with many dimensions: personal, corporate, and governmental. In this chapter we look at the dark side of the door and what might be done about it.

A brief history of hacking

Much of the activity in the early days of hacking was done for show—to gain notoriety for the hackers, mostly among hacking cliques. It's not that they didn't cause a lot of disruption and damage; rather they did it to display their cleverness, not to make money or as social protest.

Two examples among several were *Blaster*, launched in August 2003, and *Sasser*, which appeared in April 2004. Both were aimed at Microsoft Windows XP and 2000 operating systems and both were *worms*—self-replicating programs that spread on their own and carry malicious software with them.

Blaster was created by *Xfocus*, a Chinese hacker group. Later, 18-year-old Jeffrey Lee Parson created what was called a B variant. He was subsequently arrested.[3]

Sasser has been attributed to a German computer science student, Sven Jaschan, another 18 year old. He also was arrested, but tried as a juvenile because he wrote the virus before he was 18. He received a 21-month suspended sentence.[4]

The next wave of hackers had financial gain in mind. Typically they used *botnets*—small programs that infect large numbers of networked computers, mostly those with Internet connections. Botnets enable the hackers to operate those computers remotely. Typically, owners of the infected computers are unaware that their machines are compromised. Money was made from spam blasts. In addition, denial of service attacks that crash company Web site servers by overwhelming them were launched.

Hacker used to mean a person who was particularly good at programming computers and so could devise clever solutions, called hacks, to programming problems. That changed when the term became associated with people who break into computer systems, whether for "fun" or with malicious intent.

Companies may employ "good" hackers to probe their systems for weaknesses so they can shore up their defenses, but "bad" hackers abound.

Jeanson Ancheta has been credited as the first person to create a significant malicious botnet, in 2004. An American who was 19 at the time, his twist was advertising that his botnet, estimated to have control of between 400,000 and 500,000 computers, was available for use by others. The ads noted a fee scale geared to how many of the botnet's computers were to be used. Ancheta, who made about $3,000 from botnet sales and $60,000 from spam adware, was arrested in 2005, sentenced to five years in prison, and fined $15,000.[5]

As they stepped up their attacks, hackers went underground to avoid being caught. Launching all sorts of malware, some made millions by blackmailing organizations, demanding payment in exchange for not attacking their systems—the digital version of the age-old protection racket. Others made their fortunes by stealing personal data and invading banking systems to transfer money to their accounts. In one case, three young Russians, Ivan Maksakov, Alexander Petro, and Denis Stepanov, were accused of a £2 million extortion scheme against London bookmakers. They were arrested and sentenced to eight years in prison.[6]

Most attempted attacks are stopped before they can take effect. An example of one that would have set a very high bar for hackers to aim at had it succeeded, came from Yaron Bolondi. Using a *Trojan virus*,[7] he was well on his way to stealing £220 million from the London branch of the Sumitomo Mitsui Financial Group Inc. He was arrested in Israel in March 2005 after Israeli police and British officers thwarted his attempt.

Ancheta's scheme has been ramped up lately by a number of hackers who offer to sell and trade their services with each other and to potential criminals. They have created an underground market in malware, turning hacking into an ongoing business. Attacks and schemes to profit from malware have continued to grow. Automated attacks, requiring little attention or effort, have been introduced as well.

What do you think of this? *Anonymous, the once and future activists*

The amorphous group of online activists—some would say anarchists—known as Anonymous was formed in 2003 as an offshoot of posts on 4chan, a Web site known for its freewheeling and often radical content. A loosely organized cooperative community professing a social agenda increasingly aimed at championing free speech and free flow of information on the Internet, Anonymous has taken to hacking to get its message across.

Anonymous's tactics include mounting distributed denial of service attacks, defacing and spoofing Web sites, leaking documents, and posting personal information of those deemed antithetical to their causes. Among their targets have been: governments, agencies and organizations of several countries, especially those opposed to WikiLeaks; the Church of Scientology for content censorship; media and entertainment companies and their leaders, particularly those it accused of preventing free access to material; and, most recently, corporate, government, and individual supporters of the proposed Stop Online Piracy Act (SOPA).

Is this appropriate social activism and protest, or attempts to negate the free speech rights and intellectual property protection of those whose ideas they dislike?

As hacking advances, so does surveillance and development of countermeasures, one more cat and mouse game between the good guys and the bad guys. The most recent and, because of potential impact the most worrisome, is cyber warfare—using IT tools to exploit vulnerabilities in vital digitally controlled systems so as to change their performance or bring them down. The trend towards digitization combined with connectivity has left few systems untouched, hence potentially vulnerable— digital systems are overwhelmingly network-connected, so they are theoretically within reach of anyone with sufficient knowhow.

In the past, severely weakening a nation or bringing it down has been a matter of physical force, of military prowess and power. The digital revolution has opened a new avenue of attack. It no longer takes an army, just a few people with the right computer skills and a connection to the Internet. No shots need be fired, no missiles launched, no explosives planted, no bombs dropped. The new weapon of choice is the bit bomb.

Clicks on a keyboard suffice to severely cripple a country by disrupting vital digitally controlled systems—undermining the functions of electrical grids, natural gas distribution, water purification and supply, mass transit, and other infrastructure systems; shutting down or assuming control of financial systems; interrupting lines of communication.

Cyber attacks rising to the level of warfare have caught the attention of many countries. They have little choice but to build defenses against possible strikes. At the same time, they must build their own offensive capabilities to create the threat of countervailing force. While rogue hackers can mount a cyber attack, most of the danger, hence most of the action, is on national governmental levels. So except for the weapons used and the number of people required, cyber armies are built just as were old-style armies, by governments and geared for defense and offense.

That is taking us way beyond cat and mouse. We look at cyber warfare more closely farther along in the chapter.

A broad range of activities

> *. . . in the dynamic of cyberspace, the technology balance right now favors malicious actors rather than legal actors, and is likely to continue that way for quite some time.*[8]

There have always been hucksters, scammers, and snake oil salesmen, burglars, bandits, and thieves, spies, invaders, and destroyers. But there haven't always been computers and the Internet. What a difference that has made, turning crime into cybercrime—criminal activity that uses computers and digital communications systems to target computers, computer-based systems, and the people who use them.

We can be subjected to annoying amounts of unsolicited emails, promotions and scams. We can fall victim to nasty viruses that can corrupt our files and to fishing attacks that fool us into revealing personal information. Our computers can be hijacked without our knowledge, used to join in spam floods and coordinated denial of service attacks against company Web sites.

People do these things for different reasons. Some invade systems just to see if they can. Many aim at victimizing organizations. Those seeking money download credit card numbers, bank accounts, and similar records that they use or sell to other criminals.

Whatever their intent, hackers like these make victims of us all. As we've noted, infrastructure systems for power, natural gas, and water delivery are increasingly controlled digitally, hence targets for attacks that can shut them down. Military systems, network attached and remotely controlled, are tempting targets as well. Communications of all sorts, from individuals to top level agencies, wired and wireless, can be intercepted.

Thanks to digital IT, these incursions and others can be initiated from anywhere in the world. They come from many different sources: individuals operating on their own; rogue groups; agency-based operatives; and government-supported organizations. And they have many different goals: creating mischief, spying, making money, disrupting operations, and disabling systems. So we see that hacking covers the gamut, from the merely annoying to cyber warfare, in a continuous barrage. The numbers are astounding.

- In March 2009, researchers at the University of Toronto's Munk School of Global Affairs reported that nearly 1,300 computers in 103 countries had been infiltrated. The culprits worked mostly from computers in China, though there was no clear evidence of government involvement. The research, initially requested by the offices of Tibet's Dalai Lama, led to a much broader discovery that included various government offices as well as several Tibetan exile centers.[9]
- In 2011, the US Senate experienced "tens of thousands of attempted computer hackings each month."[10]
- In June 2011, Google said that hackers from central China tried to get "into the Gmail accounts of hundreds of users, including senior U.S. government officials, Chinese activists and journalists."[11]
- In August 2011, the Chinese government responded to that claim and others like it by denying that they were behind the attacks. They added "nearly 500,000 cyberattacks were aimed at computers in China [in 2010] and almost half originated overseas."[12]

If we extrapolate conservatively from these few examples, incursion attempts worldwide have to number in the multiple millions. Countermeasures have held successful attacks to a miniscule amount compared to the volume of attempts, but one success can expose huge numbers of accounts and records, or do considerable damage. Consider these examples:

- A recent survey of IT professionals in companies ranging from under 500 to over 75,000 employees found that 59 percent of the respondents "had two or more breaches in the past 12 months" and "90% had at least one." Further, 41 percent reported costs resulting from the breaches of at least $500,000; costs for 18 percent were as much as $5 million.[13]
- Network Solutions, a domain name registrar, reported that it "was hacked and used to spread malware on anywhere up to 500,000 different domains" in 2010.[14]
- In June 2011, Citigroup's Web site was a hacker's entry point into its networks, exposing credit card holder names and account numbers. "A total of 360,083 North America Citi-branded credit card accounts were affected."[15]
- Sometime prior to October 2010, files containing at least 60,000 records with names and ID card numbers held by the University of Wisconsin, Madison, were hacked. Some of the ID numbers had social security numbers embedded in them.[16] A similar event at Ohio State University around the same time exposed identifying information of around 760,000 people. "The breach will cost the university $4 million in expenses related to investigative consulting, notification of the breach, credit security and a calling center for anyone with questions or concerns."[17]

● In June a year earlier, the Stuxnet worm was employed to penetrate the Siemens control systems of centrifuges in Iran that were being used to enrich uranium, a critical step in producing nuclear bombs. On command, malware carried by the worm triggered thousands of the centrifuges to run at such high speeds that they destroyed themselves.

Inside out—leaky barriers

Corporations and governments use many reliable methods to protect their systems and files from unauthorized external access. Except where they are poorly protected, breaking into those systems requires a significant amount of expertise. Though some successful attacks have been mounted, vastly more have been repelled.

Determined thieves in the physical world know that getting through entry barriers is not a problem when the person involved is authorized to enter—the classic inside job. Digital theft is no different. Many breaches of corporate and government data security come from disgruntled employees or those out to profit from data incursions. The more employees there are who have access to various records, the greater the potential for breaches that range from simple snooping to large-scale embezzlement.

The advantage of already being inside the barriers is enormous and tempting. Most leaks of classified company and government records come from insiders. That was how WikiLeaks got hold of over 250,000 supposedly secret US diplomatic cables.

What do you think of this? *You can't take it with you—or can you?*

Corporate and government digital data theft can be as simple as copying files to a thumb drive or attaching them to an email. The former can be prevented by configuring computers to disable USB ports, or eliminate them altogether; the latter by scanning email attachments before they can be sent.

Laptops and other portable devices with sensitive data on their drives can be lost or stolen. Data on those devices should be strongly encrypted, so that even if they fall into the wrong hands, their contents will be difficult to unravel, as long as decryption keys are secure.

Assigning and maintaining access permissions is a time-consuming, tedious job. Consequently, laxity in granting authorizations and keeping them up to date can creep in. After the fact, it's often discovered that many incursions were carried out with ease by people who had access to systems for which they had no job-related business.

Also at issue is what can be accessed in systems that employees do need. For example, employee databases have categories like name, address, phone number, social security number, tax deductions, salary, job function, employment history, benefits, and incident reports. Only a very few people need access to all of that; several only need to see some categories; most don't need to see any. Keeping access permissions up to date needs to be a constant process.

Governments tend to go overboard in categorizing documents as unavailable, with such levels as restricted, classified, confidential, eyes only, secret, top secret, and more. And once a document is categorized, it likely keeps that level forever. Such overzealous attempts at secrecy result in huge stores of information. As a general rule, the more that has to be protected the harder it is to protect. Combine that with overly broad access rights and the potential for leaks and theft rises inexorably. The lesson is clear—limit access to what is absolutely necessary for job performance, keep permissions current, make secret only what really needs to be, and routinely cull the data to see what can be declassified.

The privacy sphere deflates Purposeful intrusions are not the only way that personally identifiable information (PII) is revealed. There are many instances of PII escaping the privacy sphere. Most are the result of carelessness or lax procedures. Adding to the worrisome nature of the exposures, the fact that a breach has occurred may not be noticed for some time, or worse, not acted on promptly when discovered.

- Names and social security numbers of over 63,000 living people were inadvertently included in the Social Security Administration's publicly sold digital Death Master File. An April 2011 news item[18] related that the Administration knew about it for three years and yet continued to sell copies of the file, exposing supposedly private information.[19]
- A September 2011 news story[20] told of records of 20,000 emergency room patients at Stanford Hospital, Palo Alto, CA, that became publicly available, revealing such information as names, diagnosis codes, charges, and admission and discharge dates. In a rather bizarre sequence of events, the records moved from the hospital to a billing contractor to the Student of Fortune Web site,[21] a readily viewable commercial site. It was almost a year before the breach was discovered.
- An investigation by the *Wall Street Journal* reported in October 2010 found that many of Facebook's most popular apps were sending identifying information to "advertising and Internet tracking companies" in violation of their privacy rules. The data released by the app companies included the PII of those who had the most strict privacy settings. Many millions of Facebook members were affected.[22]
- When heads of state visit other countries, their flight routes and trip schedules are closely guarded secrets aimed at keeping them safe. Yet inadvertently or otherwise, secrets sometimes leak out. In November 2011, President Obama traveled to Japan. A Japanese air traffic controller posted the details of Air

Force One's flight plans, thus exposing the president to possible danger. The controller was also accused of posting other aircraft flights, including those of surveillance drones. He claimed he had no hidden agenda or intent to endanger anyone, but just wanted to share the information with his blog followers.

What do you think of this? *Copies of your copies, records of your records*

Copier machines with such features as two-sided and multiple copies from a single scan have hard disks to support those tasks. If information on the disks is not fully erased, their possibly sensitive data is easily retrieved.

Deleting a file on a hard disk does not remove it—it only erases the indicator of where on the disk the file is stored. Unless completely and thoroughly written over, information from the files can be recovered. So be careful about what you copy on public or office machines. Businesses and individuals alike should take appropriate measures to fully erase data on the hard disks of any machines that have them before disposal of the equipment.

A big bag of tricks

There are many different ways we can be victims of IT-supported digital incursions, ranging from the annoying to the destructive. Here are the more prevalent ones.

Spam   The Internet makes it easy to send copies of a single email to any number of people. All that's needed is their addresses. That's a very efficient way for companies to send notices of new products, news releases, and recall announcements to their customers. It's also a very efficient way for spammers to bombard us with unsolicited email, most of which is aimed at getting us to spend money on products of dubious worth or no worth at all, phony get-rich-quick schemes, miracle cures, cheap but ineffective lookalike medications, and the like.

Most email spam consists of messages in the email body itself, making them susceptible to blocking by spam filters. Because of that, a once popular image spam recently made a comeback. Spam text is delivered as an image file, making it difficult for spam filters to detect because they typically work by searching for key words or phrases. Recognizing characters presented as images requires costly, processing-intensive and not always reliable optical character recognition software. As detection of static images improved, spammers took to randomizing parts of them. That again hampered detection, since strong filtering to assure blocking spam can easily block non-spam images as well.

Though email is by far the most prevalent way to send spam, any medium capable of message broadcasting is a venue for spammers. Examples include automated phone calls (robo calls), fax blasts, and instant messaging.

Spammers get email addresses by buying or trading for address lists of other spammers and by downloading them from hacked corporate lists. Another easily accessed source is email discussion groups, whose lists are available to members. So by joining large numbers of groups, spammers can create their own lists.

Many spammers take a scattershot approach, sending their come-ons to anyone whose address they can get. Others look for lists of people in particular categories and find it worthwhile to pay for them. Assembling and marketing such lists can be very lucrative.

The math is interesting. A spammer rarely gets more that a very few responses from many thousands of emails, but those responses will bring in much more money than the pennies it costs to send all the spam. Aside from the cost of purchasing lists, spammer overhead is little more than the price of a simple computer and an Internet connection.

ISPs are not so lucky. They have to maintain the capacity to accommodate spam floods at the same time as they are the first line of defense against forwarding spam. They incur real bump-ups in operating costs as spam volume rises. On the other hand, "Spam continues to plague the Internet because a small number of large Internet Service Providers sell service knowingly to professional spammers for profit, or do nothing to prevent spammers operating from their networks."[23]

Businesses whose systems face the deluge have to deal with lost employee productivity. For those of us who receive spam, beyond the annoyance factor there's the cost of wasted time, and the chance that we may fall for a scheme.

What do you think of this? *Zombies in your life*

Zombie is the name given to a computer that has been compromised by malware that lets its operations be controlled remotely. The most widespread use of zombies is to create *botnets*—legions of zombies that can be coordinated to send huge volumes of spam email and to engage in denial of service attacks on Web sites. It's not unusual to be unaware that your computer has been compromised—it may seem to be operating normally whether or not the botnet is active.

Major spammers are not individuals working alone, but rather those using spam networks whose owners sell their services to would-be spammers. Some also run their own spamming operation, tied to fraudulent come-ons.

● A notorious example is the Russian *SpamIt* that, via *GlavMed*, constructed one of the largest botnets in the world. It was used to send huge amounts of spam for their *Canadian Pharmacy* and *US Pharmacy* fake Web sites to sell counterfeit and black market drugs, no prescription needed. SpamIt appears to be closed for now.

When operating, SpamIt supplied its affiliates with Web site templates that they could use to create fake pharmacy sites of their own. About 40 percent of sales went to the affiliates, the rest going to GlavMed. At the time, "The most successful Spamit.com affiliates raked in millions of dollars in commissions. In fact, 8 out of 10 of the top moneymakers for SpamIt earned more than $1 million in commissions from Web sites they advertised via junk e-mail."[24]

- *Bredolab*, a Russian-created botnet of about 30 million zombie computers, was a major SpamIt affiliate. Taken down in November 2010 by capture of its servers, the botnet was a fee-for-service vehicle available to cyber criminals and spammers.
- The *Rustock* botnet, another notorious spam producer, closed early in 2011after a five-year run. In its day it is thought to have been responsible for about 40 percent of daily spam.[25]

The *Spamhaus Project* is a good source of up-to-date information about spam and spammers. A non-profit international organization, the Project uncovers sources of spam, works with law enforcement, and promotes anti-spam legislation. Spamhaus also maintains several current databases of blocking lists for use by service providers, corporations, and universities to exclude spam email. In addition to news, the site keeps a list of the ten worst spammers.

(http://www.spamhaus.org/index.lasso)

Just from these few examples, we can see that the volume of spam is incredible. Weekly statistics compiled by *M86 Security Labs* show that in September 2011 over 77 percent of email was spam—think of the load that put on the networks of the Internet. The three heaviest sources, India (16.1 percent), Brazil (9.7 percent), and Russia (7.7 percent), together accounted for over a third of all spam.[26]

Denial of service attacks   One way to disrupt a company's Web site is to bombard it with such a high volume of requests that it slows to a crawl or shuts down completely. That's a *denial of service* (DoS) attack. It's cousin, *distributed denial of service* (DDoS) attack, has the same goal but is instituted as a coordinated bombardment by many computers. Often using zombies, an attack can be launched with just a single command.[27]

Neither bombardments that aim to overwhelm resources nor false authentication information require hacking into systems, so they are easier to mount. Normally, when a Web server gets a request to authenticate a user, it responds with an acknowledgment of acceptance that grants access to the site. In a DoS attack, not only are a great many user requests sent, but the acknowledgment return addresses are false, so the server can't find the users. Even though each request is discarded after a few moments, a sufficiently large volume of requests will bog down the server as it attempts to respond.

Many more involved schemes do involve hacks or spoofs. Each one exploits some system function that becomes a vulnerability. Once hacked, Web server operations can be disrupted by changing key elements of the server's operating system or configuration files.

As each type of attack is mounted, defenses are developed to thwart it, but as is typical, new attacks are continuously being developed—one more running battle. Old attack methods can work too, against any Web sites whose protections aren't up to date.

Spoofing   Filtering systems can be skirted and users fooled by making messages and Web pages look like they're coming from a legitimate source. Here are the most common of the many ways this simple trick can be used.

Email spoofing changes the "from" section of the email header to that of someone the recipient knows or to an anonymous or well-known source. Depending on intent, that can be good or bad. For example, a whistleblower might not want to be identified at the outset when sending sensitive information about company misdeeds.

More disturbing is when an email looks like it comes from a trusted source, often someone in your contact list, that encourages you to click on a link—*I have an important message for you; Look at these great pictures I took; Let's get together and catch up.* When you click, you wind up downloading malware.

Another trick is the plea. You get an email from someone you know, perhaps a friend or even a relative, who claims to be stranded somewhere without any cash or credit and needs you to wire money—only $500, or maybe $1,000. Empathizing, you do, only to learn later that the friend or relative is just a spoofed email header.

Then there is the email that looks like it comes from your bank, right down to the layout, fonts, and logos. The typical message is that there is some sort of problem with your account. All you need to do to straighten it out is click on a link that takes you to a legitimate-looking Web page asking for your account number and password so that your account can be validated. If you supply them, say goodbye to your money.

A similar way to get you to part with personal information is *Web page spoofing*. Hackers can break into corporate and government sites and replace their Web pages with others. It may be done as a practical joke or to make a point. It may be justified by the miscreant as punishment for a perceived grievance. It also may be aimed at us, to lure us to a site that looks like one we know. We go, it looks legitimate, so we log in, and the hackers are off to the races.

> JPL was founded by the California Institute of Technology (Cal Tech) in the early 1930s for rocket research and development. Federally funded, it now is an adjunct of the US National Aeronautics and Space Administration (NASA), though still managed by Cal Tech. Its latest projects involve running NASA's Deep Space Network of internationally based terrestrial antennas, building and operating robotic spacecraft, and supporting earth orbital missions.
>
> (http://www.jpl.nasa.gov/)

- Early in 2003, several Jet Propulsion Laboratory (JPL) Web site pages were replaced with pages containing messages protesting US policies towards Iraq and the imminent threat of invasion. The replacements were the work of the so-called *Trippin Smurfs* hackers, who broke into JPL's servers.[28]
- In May 2011, *LulzSec* hacked into Public Broadcasting System (PBS) servers to post a phony article on the PBS Web site about slain rapper Tupac Shakur, claiming that he was alive and living in New Zealand. While they were at it, they stole PBS staff email addresses and passwords, posting them on LulzSec's Twitter pages. Apparently it all was in reaction to negative publicity about WikiLeaks.

What do you think of this? *Heroes or villains?*

Lulz Security, aka LulzSec, was a group considered by many to be six ill-intentioned hackers out to do varying degrees of damage. The group is variously depicted as pranksters, practical jokers, disrupters, and cyber terrorists, the latter coming from the Arizona Department of Public Safety, one of their targets. Other victims include Sony, Fox.com, News Corp, Nintendo, and the CIA. The group also has been credited with rousing a wave of imitators as well as groups that have gone after LulzSec itself.

Some people commended the group for providing a public service by pointing out security weaknesses in different Web sites and attached databases. Pointing out involved hacking into the sites, replacing Web pages with various messages, humorous and not, and stealing data. After 50 days of operation, they declared in June 2011 that they were shutting down. (For the text of their "final release" see: http://pastebin.com/1znEGmHa) Whether any of their members pop up in other guises, or the group resurrects, remains to be seen.

IP address spoofing tricks systems. Every device attached to the Internet has an IP address that identifies the network the device is in and ultimately the device. The source and destination IP addresses of transmitted information are part of the header of the data that make up the transmission. The Internet's routers use the destination address to direct the data through the various links that form a path to the intended receiver. Security mechanisms check incoming source IP addresses to block those from sites or systems known to be sources of malware and attacks. By changing those source addresses to benign ones, security can be bypassed.

Telephone systems are not immune. *Caller ID spoofing* makes an incoming call look like it's from someone other than the actual calling party. What shows up as the caller can be a spoofed phone number or name. You will be more likely to answer a call from a number or name you recognize, only to be treated to a solicitation or worse—an attempt to get you to reveal personal information. That leads us to the next trick.

Social engineering A surprisingly effective way to get people to reveal passwords, account numbers, and the like, or to click on links that transmit malware or connect to fake Web sites, are techniques that fool us into thinking they are just what they appear to be, though they are not. These furtive approaches rely on pulling the wool over our eyes, akin to the con artists and flimflam men of yesteryear.[29]

- A popup warns us that our computer has just been compromised by a virus and to click on the link to eradicate it.
- An email from a bank asks us to validate our account by logging in.
- An alert from a credit card company tells us our card may have been used by an unauthorized person and to click on a link to check the purchase.

- Emails disguised as fun quizzes, simple contests, product giveaways, surveys, and the like, capture information required to be entered before proceeding.
- Telephone calls asking for identifying information seem to come from legitimate financial institutions.

These social engineering techniques, called *phishing*, are analogous to casting a broad net to catch fish. The hope of the social engineers is that some of the many who are phished will bite. In the midst of concentrating on a multitude of activities, we may be distracted enough to respond without checking validity. If enough personal information is gathered, identity theft follows.

As we've become more aware of phishing attempts and filters have gotten better at blocking them, a more targeted version called *spear phishing* has been employed. Instead of the indiscriminate barrage approach, spear phishing focuses on select groups that the phishers know something about. Particular employees get email that appears to come from a company's vendors. Members of a professional group are offered discount insurance designed especially for the group. People who subscribe to certain magazines are asked to participate in a survey. The aim is the same as general phishing—to get access codes, account data, other personal information, or to transmit malware.

A recent venue for these attacks forsakes conventional email, instead using social networking sites. With hundreds of millions of participants, social networks are extremely tempting targets of easily and precisely focused attacks. The same psychology applies—when messages seem to come from friends, they are more likely to be trusted. So spoofing and phishing combine to get at the mountains of personal information supplied by social networkers, for all the same nefarious reasons. You can be victimized in two ways—falling for a scam or being used without your knowledge to induce others to bite.

What do you think of this? *Who are you?*

Using certain pieces of information, like Social Security or driver's license numbers, your identity can be assumed by someone else. Then false credentials can be created, credit card accounts opened, bank accounts accessed, and so on. As each credential and account is created, it becomes another identifier to help in creating more.

Since innumerable purchase and financial transactions are carried out online, the thief doesn't have to bear any resemblance to you and you will have a hard time proving it wasn't you. If questioned or caught, criminals can use falsified credentials to identify themselves to police, potentially saddling you with a criminal record. Debt run up on fraudulent cards can result in your being sought by collection agencies. Your credit rating can be demolished and it can take enormous effort and a long time to get things straightened out.

Malware Viruses, Trojans, worms, spyware, keyloggers, rootkits, and URL and SQL injectors are among the most common categories of malware—software designed to do bad things to computers, computer-based systems, and unsuspecting victims.

New malware programs are created every day and there are many varieties of the same type of malware in every category. Malware creators are a very clever, albeit ill-intentioned, lot. Many of them post their constructs online, available for anyone to use. Anti-malware producers work hard to keep on top of the onslaught, but until updates are developed and incorporated into our protective software, some malware can get through. Without going into technical details, let's see how they do their dirty work.

Viruses These are malicious software that usually arrive at your computer by hiding in email attachments, something you ask for on a Web site, files you download online, or from programs on thumb drives or CDs. When you open the attachment, click on a link to the Web site, or open the file, the virus program runs, infecting your machine. That can be as annoyingly innocuous as putting strange messages on your screen or as nasty as deleting files on your drives or altering critical startup code to render your computer unusable.

Aside from what viruses may do to a computer, what makes them so malicious is that they duplicate themselves and transmit the copies to computers you contact, usually by hiding in email attachments. Some viruses can use your email contacts list to send an infected file they've brought with them—hidden in a joke, a photo, a link—to everyone on the list without your knowledge or participation. As soon as the attachment is opened, the virus is activated, the target machine is infected, and the infection spreads.

> Trojans, also called Trojan horses, get their name from the myth of the Trojan War, wherein the Greek army left their enemy city, Troy, a giant wooden horse as a parting gift, supposedly signaling the end of a nine-year war. The Trojans took the horse inside their fortifications, not knowing that Greek soldiers were hidden inside. These soldiers emerged at night and conquered the unsuspecting Trojans.

Trojans Carrying their cargo in disguise, *Trojans* can look like useful software or files coming from trusted sources, but looks aren't everything. That cute little free game, an offer for a discount on a popular product, an app to clean your hard drive, a tasty looking recipe, or voyeuristic photos of celebrities, may bring along a Trojan. Some are just mischievous, but most are malicious. The worst kind can open hidden entry routes to your computer—so-called backdoors—that give hackers access to your machine. Then your files can be read, your data and personal information stolen, and your computer turned into a zombie. What Trojans can't do is duplicate themselves or infect other files.

Worms These are like viruses in that they can reproduce themselves and travel from computer to computer, infecting them as they go. But worms are worse, because unlike viruses, they don't need your help. On their own they can make any number of copies and spread via your contact lists and by jumping onto files that you send, all without your awareness. Actions are similar to viruses—damaging files and programs and opening backdoors.

Spyware   As its name implies, *spyware* collects information about what we do on our computers without revealing its presence. Their activity can slow down the machines, though not always in a readily perceptible way.

It is not unusual to find ill-intentioned spyware installed on machines in public places like Internet cafes. Using those machines to check bank balances or make an online purchase with a credit card, for example, will result in your account numbers, card numbers, and passwords being stolen. Some corporations use spyware to check on the online activities of employees. Privacy and ethical considerations jump into the picture.

Adware, software that pops up advertisements, is considered to be a type of spyware if the ads it shows are based on what it finds you're doing on your machine at the moment. This is different from targeted advertising, which constructs its targets based on information gathered from histories of product searches and purchases on visited Web sites.

That is considered to be a minor distinction by some, because in either case, ads are displayed in response to data collected largely without user knowledge. Adware is less ethical, however—most Web sites provide information about what they collect and what they do with it, although they don't always make that information easy to find and we seldom read through those voluminous disclosures. Spyware, by contrast, always operates in stealth mode.

Spyware doesn't duplicate itself or spread the way viruses and worms do. Most of the time it arrives on Trojans or attached to downloads from less than forthright sites and from freeware offering games, computer cleanup programs, speedup utilities, and the like—typical social engineering techniques. Spyware also can be installed by worms and viruses, but that's not the usual route.

Keyloggers   Actually a form of spyware, *keyloggers* record every keystroke you make. Some also can capture screenshots. Basic loggers save the records on your computer's hard disk, from which they can be retrieved remotely by software similar to the kind that turns computers into zombies. Other loggers transmit their collected data automatically to collector sites when you are online. Either way, as with other kinds of spyware, you are not aware that it's happening.

Rootkits   Created to help network administrators and software programmers remotely update and manage computers, *rootkits* have become a dangerous tool in the hands of hackers. Among the nastiest of malware, rootkits embed themselves deeply into computer operating systems. Then they can provide remote entry to stored data, the system and configuration files that control the operation of the machine itself, and any installed software.

Strictly speaking, rootkits don't do any damage themselves. They do their dirty work by turning over the operation of your machine to remote operators. That's the main event and their greatest danger. By giving remote control to people on command, rootkit-infected computers can be used to create all sorts of mayhem.

Rootkits arrive at your machine the same way as other malware, but aspire to a higher level of treachery. By burrowing into a computer's operating system, they run at a level of computer control that supersedes almost all the other programs on the machine, the same as do most antivirus programs. That makes them much more difficult to weed out once they take hold.

What do you think of this? *You bought a rootkit!*

In 2005, Sony BMG, in a move to enhance copy protection on its CDs, added two programs, one they called *Extended Copy Protection* and the other, *MediaMax*. The former was placed on 52 titles and the latter on another 50. Play one of those CDs on your computer and the programs automatically installed.

These programs actually are rootkits. They did enhance copy protection for Sony BMG, but they also exposed computers to easy exploitation by hackers. In the uproar that followed, including a number of lawsuits, the company recalled all the affected CDs. Cleaning infected computers was another matter, though. One good outcome—attention was drawn to company practices that install spyware and rootkits to unsuspecting users' computers. Was it enough?

Injectors   Type a uniform resource locator (URL) into the browser address bar and the URL is switched to that of an affiliate link, taking you to a different Web site than the one you sought. That's a *URL injector* in action. It may be used to bring payments to the injector supplier for redirecting users to a particular site. It may send you to a spoofed site that tricks you into revealing personal information or clicking on a link that downloads malware. Any way you look at it, your browser was hijacked.

Code injectors insert programs that affect the operation of the machine. That's one of the ways worms spread from computer to computer.

SQL injectors are the most common malware behind database break-ins. By using the language of database queries (SQL), they gain access to the records themselves. These injectors also can modify SQL queries to change logins and user names, thus clearing the way for unauthorized access. Injector-based attacks can operate automatically or be manually controlled.

Fighting back

Malware Protection is No Longer an Option—It's a Necessity

- *90% of new malware is designed to infect users through the Web*
- *Up to 85% of all Web-based infections occur through legitimate Web sites*
- *Six in 10 malicious URLs pass undetected through most Web security solutions.*[30]

The Internet and the Web are not the only sources of malware, but they are such overwhelming sources that they deserve special attention. We've seen the most prevalent types of malware, the routes they usually take, and the damage they can do. The best way to avoid those consequences is to stop malware before it can be installed. That means effective firewalls, proxy servers, antivirus programs, and diligence in choosing the sites and files we access and copy.

Firewalls   These are hardware and software working in tandem to create entry barriers to malware and non-trusted sites. In corporations, they sit at the point where corporate networks attach to the Internet. Incoming communications must pass through the firewall to reach the enterprise networks.

We can install firewalls in our personal computers—their operating systems typically include firewall software, as do most antivirus programs. In effect, the firewall is interposed between our computer and the connection to our ISP.

Firewalls work in a number of ways, but all have the same purpose—to screen out malicious attacks. While they can be highly effective, they are not perfect. Furthermore, the more thoroughly they examine incoming transmissions the more they slow down the connection and the more likely they are to block content that should be allowed to pass. So there's another balancing act to perform.

Proxy servers   Also hardware and software combinations, proxy servers are used in businesses to add another layer of protection by acting as intermediary relay stations. Instead of a direct connection between company Web servers and the Internet, requests and responses go through the proxy, which decides whether or not to pass the information.

For example, an external request requiring a response from a company server first goes to the proxy. Should it be deemed safe, the proxy retrieves the appropriate response from the company server and relays it to the requester. So there is no direct connection between the external request and the server. Similarly, outgoing requests are evaluated by the proxy server and are relayed to the requested site if considered legitimate.

Proxy servers often are set up in conjunction with firewalls and may be considered to be a firewall extension. Increasingly popular for corporate resource protection, they are too elaborate for individual use.

Screening software   As well as businesses, we can also benefit from software that screens incoming downloads and files to block those carrying malware. Installing such programs on individual computers is a good idea. Corporations are likely to select one vendor's product and install it on all the enterprise's machines. For our own use, there are many anti-malware programs available. Some have free versions of their more elaborate offerings and others have one-time or ongoing fees.

Aside from virus monitoring of downloads and file transfers, they may have features like email checking, spam and adware filtering, spyware capture, and even rootkit blocking. They also can remove malware that has made its way past the barricades, next best to stopping it beforehand.

Antivirus programs   There are a lot of applications to help us clean out computer viruses. They work by building databases of malware *signatures*, telltale bits of program code that reveal them to be malware. Once recognized, viruses are removed or quarantined—placed in hard disk locations from which programs can't execute. The most effective programs add real-time scanning of incoming material to user-invoked scans of entire disks. Scans may be set to run at specific times or started manually.

Some programs incorporate heuristics that attempt to identify viruses in other ways than signatures. Though not perfect, they can be effective in catching new viruses whose signatures are not yet known. They may deconstruct a suspect program and compare its code with known virus-like sequences; they may run the software in a protected area of the computer to see what it tries to do. Anything suspicious is quarantined unless you indicate otherwise.

How well antivirus programs work depends not only on the quality and features of their designs, but most importantly how current their signature databases and detection engines are. Some vendors provide daily database updates, others less frequently. Some programs can be set to update automatically, others require manual updates. The same applies to detection engine updates.

Spam avoidance   Spam suppression is mostly the province of ISPs and email systems operators. They build tables of known spammers and characteristics of spam mail based on subjects and a variety of suspicious words. Incoming email is filtered out if it fits the descriptors. Email programs can also block senders whose email we mark as junk. If we receive spam that the filters miss and declare it to be spam, junk mail, or as coming from undesired sources, the tables are updated and the filtering improves.

As spam filter performance advances, spammers try ever more sophisticated ways to get into our computers, so a lot still gets through. That has led to a proactive approach that doesn't rely on filtering. The idea is to block the ways spammers make money. Products sold via spam are paid for by credit and debit cards, so if the banks issuing those cards refuse to process the payments, spam loses its point. Furthermore, it doesn't matter where the spammers are located since there is no need to find them.

Of course, it's not that simple. First there is the legality of refusing to process a charge instituted by a valid card holder, regardless of what's being purchased. Instead of automatically rejecting a charge, card holders could agree to receive alerts when a product offer is from a known fraudulent spammer and have the option to cancel or go ahead with the purchase.

Card issuers would have to be certain that the purchase actually was spam related before sending an alert, or possibly face lawsuits. Then too, card issuers don't make money by refusing transactions, so their incentive would have to be acting ethically, which they could capitalize on by making it part of their offerings package to attract people to apply for their cards.

Yet another approach is to go after the companies whose products the spammers are selling. That's also not so simple. While offers are likely to be for knockoffs, bogus medications and potions, useless sex aids, health-related devices touted by unproven claims, and all sorts of non-functional gadgets—clearly fraudulent—they can be produced anywhere in the world, beyond the reach of a nation's laws.

The US government stepped in with an effort called the *CAN-SPAM Act of 2003*, which took effect in January 2004. Its actual name is the unwieldy Controlling the Assault of Non-Solicited Pornography And Marketing Act of 2003. Its aim was to reduce spam by establishing rules that legitimate commercial emailers had to follow. Those that do not are considered to be illegitimate spammers, subject to significant fines. The Act is enforced by the Federal Trade Commission and the Federal Communications Commission. The latter was required to create rules for wireless transmissions.

The major CAN-SPAM rules[31]:

- The sender must be accurately identified. All from and to lines have to have correct, non-spoofed email addresses and header routing information.
- Email subject lines have to clearly indicate what the email is about.
- Recipients have to be able to unsubscribe from the list.
- An actual physical business address has to be part of any commercial email.

For the Act to be effective, it must be rigorously enforced and action taken against violators. That's easier said than done, especially because much spam originates offshore from locations not subject to jurisdictions cooperating with US regulations.

Identification   Organizations need to know that logins to their Web sites are legitimate, so some means of positive identification is needed. There are three broad categories of identification methods: things we know, like passwords and personal identification numbers (PINs); things we have, like identity (ID) cards or swipes; things native to us, like fingerprints and retinal patterns.

Ordinary identification by user name is simple but not, by itself, reliable. Most systems will remember your user name so you don't have to enter it each time. Of course, that can reveal the name to anyone nearby. Still, name is rarely enough to get into a system.

The next step is verification, which typically means passwords and possibly extra security questions—your mother's maiden name, your first pet's name, the grade school you went to, where you were born. For secure transmissions, more complex schemes are required, most of which involve exchanges of encrypted information.

Passwords, paradoxical   A simple password—a name, a city, an auto brand, a word in the dictionary—is easy to remember and easy to figure out. Adding a few numbers or mixing in lower and upper case letters doesn't help much. Computer programs can run through those kinds of passwords in seconds, and if you know something about the person, guessing works surprisingly well. Furthermore, using the same password for many sites means cracking one opens all. Simple solution: use a different password for each site. Not so simple: remembering all of them.

Complex passwords—meaningless pseudo random combinations of upper and lower case letters and numbers—are harder to crack. The longer and more random they are, the less likely they can be divined and the harder they are to remember. Create a different one for each site and the difficulty soars. The odds are pretty good that you'll forget some, especially those for sites you don't often visit. Besides, we're told to change our passwords fairly often.

So, write them down—in a file on your computer that can be hacked? On a flash drive that can be lost? On a piece of paper in your desk drawer? On a note stuck to your monitor? That's the password paradox—the better and more varied the passwords, the harder they are to remember, so the more likely we are to use aids that open them to discovery.

Other forms of passwords have been tried to get around the memory problem. Many are based on biological identification—fingerprint and palm print readers, retina scanners, facial recognition software. Except for sophisticated expensive systems, these are not very reliable, especially those that are installed on microcomputers and laptops.

Another way—graphical passwords   We are better at recognizing pictures than we are at remembering words, numbers, and strings of letters and numbers. Taking advantage of that are graphical passwords. One example is a system called *Passfaces*.[32]

To set our "password" we select four faces from a large number that are shown to us. Then to log into a system, we have to click on the four we've chosen, each presented one at a time randomly placed in a different array of nine faces. The odds of someone guessing which faces we chose are quite small, but we can recognize them quite easily. Though more tedious than simply entering a password, Passfaces is much more secure while avoiding the password memory problem.

Human at the keyboard   One way for hackers to slow a company's Web site to a crawl is to have computers continuously and automatically bombard it with registration requests. To prevent that, many sites use *Captcha* as a way to know that a human is at the keyboard.[33] The application presents a distorted sequence of upper and lower case letters and numbers that must be typed in to register. People are good at reading those sequences, but computers are not, at least for now.

In the same way, Captcha can be used to verify that user ratings, blog comments—anything that is supposed to be coming from people—actually does. Captcha's job is not to authenticate users—it's to distinguish between people and computers. However, it's easily combined with user authentication techniques.

Two key user authentication adds security   Encryption, which renders files unreadable without a deciphering key, is used to protect stored and transmitted data. For it to work, the key has to be transmitted to appropriate persons without it being intercepted. A popular solution uses two keys, one call public, the other private. This system falls under the general heading of *public key infrastructure* (PKI). Here's how it works.

You want to send a secure transaction to a bank. The bank sends you a public encryption key, one that anyone can have, but keeps its private deciphering key. You encrypt your transaction with the public key and send it. Since only the bank has the private key to decipher it, even if your message is intercepted, it's unreadable without the private key.

Reversing the process can provide authentication. You send your public key to the bank and use your private key to encrypt your signature message that you send as well. The bank decrypts your signature, which must have come from you since you are the only one with the private key that encrypted it. Of course, anyone who gets your public key could read your message, but all that would do is let them know that you sent it.

Putting the two together provides both secure transmission and authentication. First, encrypt your message with the bank's public key. Then encrypt it again with your private key. The bank uses your public key to decrypt the transmission, which authenticates you, and then uses its private key to decrypt again, revealing the message itself.

That sounds good, but where do those keys come from and how do we know they are valid—actually from the organization they seem to be? Businesses using PKIs commonly rely on a *certificate authority* (CA), an organization that has the trust of the parties at both ends of a transaction. The CA issues a *digital certificate* to the business, which certifies that the named holder of the certificate is the one that owns the public key.[34] That verifies that information sent to it by the associated private key is actually going to the source it claims to be, not a spoofed site.

The CA uses its own private key to digitally sign and publish the public key that's sent to the user. That process, which associates the public key with a particular owner-entity, is called *binding*. The trust placed in the CA by both ends of the transaction relies on their belief that the CA's private key is valid and held securely. If it's not, the whole process falls apart.

Another identification system is called *two factor authentication* (TFA). To log into a system, you must use two different forms of identification from the three possible groups—something you know, have, or is part of you. RSA, one of the most popular providers of "security, risk and compliance solutions," issues digital certificates for PKI and also tokens for TFA, which they call *SecureID*. Tokens are the additional item of identification that TFAs can use to assure logins are authorized. They can be generated by a small thumb-size plugin device or by software, and can include digital signatures.

What do you think of this? *Tokens, tokens, who's got the tokens?*

In March 2011, RSA's networks were invaded. Stolen seed values that SecureID uses to generate tokens could be used to generate fake tokens for specific companies, exposing their systems to unauthorized logins. Program code theft could be used to find private keys to access RSA's own servers and tokens.

Altogether, as many as 40 million corporate and government employees could have been affected. (Source: http://venturebeat.com/2011/03/18/rsa-security-breach/) Gartner estimated that RSA provided tokens for about half of the 80 percent of US banks that use them. The breach could cost the banks as much as $100 million to replace the compromised tokens. (Source: http://www.bloomberg.com/news/2011-06-08/emc-s-rsa-security-breach-may-cost-bank-customers-100-million.html) The issue confronted by IT specialists is whether to continue with token-based TFA or switch to something else. The problem is finding something else that is any better.

Code signing   A type of digital signature for software downloads, code signing verifies that the software has not been altered since it was signed and that its publisher is a trusted party. Updates and subsequent new version releases, when code signed, let the downloader know that they are not false releases set up by hackers. Encryption keys are used to form the code signature, as in PKI systems. These are most often managed by certificate authorities. Software publishers can create their own keys as well.

Securing transmissions   The Internet is an inherently unsecure communications system. In its original design, reliability was the prime goal—get the message through no matter what was going on in the networks and systems that comprise the Internet. There was no hint of the enormous abuses that came to be quite an unintended consequence.

The Internet remains unsecure, but methods have been devised to layer security onto the transmissions. One of the most popular is called *tunneling*, which uses various protocols to hide messages in a sort of encrypted wrapper. Businesses can use tunneling software to create secure paths through the Internet, but if the ends of the paths are not secure, they are open to invasion. Conversely, if the ends are secured but the path isn't, then transmissions are vulnerable to snooping without the need to hack into end systems.

Hacking the endpoints can yield entry into corporate and personal systems, exposing all sorts of information to theft and corruption. Following Internet links to end system computers is the way to find those points. Getting past security barriers is the next step. Hackers have some advantage there, because barriers against unknown forms of attack are only partially successful. It's only after an attack occurs and is dissected that a comprehensive defense can be employed. That emphasizes the point that entry barriers alone are not sufficient—sensitive data should be strongly encrypted as well.

Honeypots—laying a trap to catch hackers   Since hackers can be anywhere in the world, can easily move to different access points, and can hide behind anonymity, finding them is not easy. Luring them into a trap is one way around that dilemma.

A *honeypot* is a computer that seems to be a vulnerable part of a network or linked to a server or other resource that has valuable information. That makes it a tempting target for hackers. In fact, the honeypot is not used for anything else in the organization, contains nothing of real value, is isolated from the company networks, and is closely monitored. Since it has no internal use, any activity on the machine has to come from probes or attempted attacks. They are logged and analyzed to find the hackers.

Cyber terrorism and cyber warfare—a rain of bit bombs to attack people, systems, and infrastructure

We worried for decades about WMDs—Weapons of Mass Destruction. Now it is time to worry about a new kind of WMDs—Weapons of Mass Disruption.[35]

The difference between cybercrime, cyber-espionage, and cyberwar is a couple of keystrokes. The same technique that gets you in to steal money, patented blueprint information or chemical formulas is the same technique that a nation-state would use to get in and destroy things.[36]

Much of what developed countries and their populations rely on for everything from public utilities to government operations depends on digital communications and digital control. Ratchet up malware targets to those arenas and we have *cyber terrorism*, the makings of *cyber warfare*. In contrast to terrorism that relies on suicide bombers, car bombs, rocket attacks, and crashing planes into buildings, and to warfare that requires armed forces and huge amounts of equipment and money, cyber attacks take place over communications systems, primarily the Internet. A rain of bit bombs can be launched by anyone with a connection, from anywhere.

Individual rogues and small hacker groups have hit some systems, but the real threat comes from government-supported efforts to create methods of attack that can do substantial, even crippling damage. Exploiting a country's digital vulnerabilities is the key. Aims range from espionage, which could be conducted on a large scale by botnet invasion, to infrastructure shutdown.

What do you think of this? *A call to arms?*

The hacker group Lulzsec, which we've seen has been responsible for many hacks into commercial and government systems, posted "Operation Anti-Security" on June 19, 2011, saying in part,

> . . . government and whitehat (*sic*) security terrorists across the world continue to dominate and control our Internet ocean. [We are] declaring immediate and unremitting war on the freedom-snatching moderators of 2011.
>
> . . .
>
> . . . we encourage [anyone] to open fire on any government or agency that crosses their path . . . Top priority is to steal and leak any classified government information, including email spools and documentation. Prime targets are banks and other high-ranking establishments.
>
> (http://pastebin.com/9KyA0E5v)

Points of attack All sorts of a country's vital systems are controlled by network-connected computers. Infrastructure distribution systems for electricity, natural gas, and water increasingly are managed by computer-based digital controllers. These are connected to the systems by communications networks so they can be operated remotely. The right hack can gain entry into those networks and through them to the controllers. Then their operations can be disrupted or taken over, directly affecting connected systems.

The results could be massive blackouts, disruption of gas and water delivery, manipulation of water purification equipment to let untreated water flow through distribution pipelines, and shutdown of oil refinery operations. Air traffic control and guidance, financial systems of all sorts, train and other mass transit services, telephone call routing, and much more could be undermined. The more that networked digital control becomes the norm, the greater are the opportunities for sabotage.

Organizations also can be the targets of attacks. When their operations are woven into the daily patterns of everyday life, compromising them can be quite destabilizing. ATMs could stop working or could spew out all their cash. Credit cards could be deactivated or huge charges posted to accounts. Package deliveries could be rerouted. Opportunities are limited only by imagination.

Government agencies and the armed forces are not exempt either. Agency email systems could be hacked. Web sites could be flooded with bogus transmissions or replaced altogether. Sophisticated malware can interrupt military functions. Secret information concerning global affairs, military maneuvers and capabilities, and cyber defense systems could be stolen.

Just as anti-malware programs are needed to protect computers, so anti-cyber attack mechanisms are needed to protect nations and their populaces. In yet another cat and mouse game, though one with extremely serious consequences, offensive and defensive bit-based weapons and counter-weapons race to gain an advantage. All developed countries are taking steps to defend against cyber attacks at the same time as they are creating attack modalities and capabilities of their own.

In a sense, the scenario is an alarming flashback to the cold war, where the threat of mutual annihilation from a nuclear onslaught kept the war from turning hot and destroying life on our planet, though a few times it came very close to just that. We could hope that the threat of mutual annihilation by massive bit bomb attacks to carry the same doomsday scenario would keep it from actually happening as well. But there is a gigantic difference from the old cold war era.

Waging nuclear war required enormous investments, scientific and engineering knowhow to build plants to create fissionable material and build bombs, construction of thousands of planes and missiles to carry them to their targets, development of complex tracking systems to warn of attacks and quickly launch counterattacks, intricate guidance systems to control offensive actions, and much more.

Now we know that little beyond some very clever programming developed by small cadres of people under government guidance can launch an attack. And instead of having a few hours after missile launch to react, it's now just seconds. That means that defensive systems have to be automated—computer controlled. In large measure, the conduct of cyber war would have to be turned over to machines.

Software to counter and engage in cyber warfare, necessarily extremely complex, would be prone to error and unforeseen chains of action/reaction. A small programming bug or hardware glitch could mean failure to defend in time or an attack launched in error. The more these systems are automated, the greater the risk for disastrous unintended consequences. Must we go back to building bomb shelters? Bit bomb shelters are not so easy to construct.

What do you think of this? *Stealth destruction by remote control revisited*

We've seen that the Stuxnet worm was specifically designed to carry malware that instructed Siemens programmable controllers to cause centrifuges in Iran used to enrich uranium, possibly for use in atomic weapons, to spin so fast that they destructed.

The most chilling part of the story is that programmable computerized controllers are commonly used to run industrial machinery of all sorts worldwide, leaving them vulnerable to cyber attacks that could cause widespread damage. The consequences for businesses, industries, and nations can be huge.

And so—keeping the bad guys at bay

Distrust and caution are the parents of security.[37]

In every era, the IT of the time was available for use by anyone with access, regardless of their intentions. So it is now, but with a big difference. The vast array of computing equipment, communications infrastructure, and software that is today's IT brings with it the enormous multiplier effect of the Internet.

The massive amount of valuable reliable information reachable by the Internet is far beyond precedent, and so is the huge amount of unreliable, falsified, spurious, and mal-intentioned information. The availability, global reach, numbers of victims, and scope of potential damage from even a single malicious outbreak is unparalleled.

Heightened security is vital to prevent incursions of all sorts and to counteract them once they have occurred. We know that measures ranging from installation of antivirus software to intense scrutiny of transmissions and multistep sophisticated authentication techniques are needed. But we also know that without our active participation, that only gets us so far.

We've seen that intrusions can be quite small and relatively harmless, massive and potentially devastating, and everything in between. The one constant is that attacks can't take place without access to their targets, so controlling access is the key to repelling intrusions and securing systems. Once access is compromised, damage control must take over to limit loss, restore functioning, and dispatch intruders.

The overwhelming majority of bit attacks take place via the Internet. All of us, individuals and organizations alike, are outside the Internet. We connect to it through access providers, so it would seem logical that those entry points are the places to erect barriers, to exercise control over what passes to us and what doesn't, what can be accessed and what can't. The problem is, that sort of control is anathema to most of the world's more enlightened societies. Experience has shown that even in those nations that impose strict limits on access, sometimes going as far as shutdowns, workarounds can reach many sites and communications channels officially off-limits.

More immediate points of entry for us are our own computers, and for organizations, at their edge routers—the points where their internal networks connect to the Internet. That's where firewalls and filtering software are located, where a great deal of the action takes place, and where intrusion prevention is practical.

As we've also seen, how well that works depends as much on us as on technology, whatever devices we're using—our own, those of our employers, library computers, and so on. We must be on the alert for phishing scams and other forms of social engineering, keep anti-malware software up to date, ratchet up privacy settings on social network sites, and be circumspect about the sites we seek, the links we click, and the attachments we download.

Hackers gravitate towards the most lucrative and most widespread targets—so-called high value objectives. A fast-rising example is the business trend to incorporate Web-based online applications in their operations. Those present easy targets for hackers to find and steal corporate and private data. The same applies to government operations and their Web sites as well.

Consider the points of entry that can be exposed in a chain of events that characterize many networked transactions systems. A prime case in point is credit card processing. From the time you swipe a card until your account is charged, the merchant is paid, and you make a payment on your account, many steps take place. Almost all of them are network-based and outsourced, opening many potential points of vulnerability.

When outsourcing involves other nations and jurisdictions, the piecemeal and contradictory collection of data security legislation and regulating authorities that exist further complicate the picture. Hackers don't hesitate to take advantage of the situation and the chance to capture many thousands of credit card numbers, a highly saleable commodity in online marketplaces specially set up by and for criminals.

A rising challenge to us and organizations of all types comes from the growth in wireless communications, especially mobile devices. As more and more intelligence is built into them, these everyday products are beginning to attract hackers who see them as inviting targets. Smartphones used for banking, as charge cards, mobile wallets, airplane boarding passes, and more will be tempting. As yet, few hacks have been reported even though protection for those devices is quite limited. But as their functions and popularity expand, that's sure to change.

What do you think of this? *Does smarter mean more vulnerable?*

It is becoming common for automobiles to be equipped with factory installed antitheft systems, keyless entry, keyless starting, and GPS navigation. Many of those systems are connected wirelessly to cell phone, WiFi, and other similar communications networks so that remote services can be provided on request. All of that makes them potentially vulnerable to hacks that call those systems into action, subverting their intended use. Hackers can tap into them to send signals to unlock doors and start a vehicle, to capture stored GPS routes, to listen to phone calls, and to disable a vehicle altogether.

Even simpler—if your car is broken into, the miscreant can click the "home" destination on your portable or built-in GPS device and follow it to your abode, which may be a tempting target for a burglary or home invasion.

Security adds other issues

If you think technology can solve your security problems, then you don't understand the problems and you don't understand the technology.[38]

The need for security is undeniable, but it also is used as justification, or should we say an excuse, for invasions of privacy that otherwise would be out of bounds. Stopping malware, blunting spam, and securing data are crucial, but so are preserving privacy, individual freedoms, and access privileges. Where to place the balance point depends on the extent of potential threats and potential damage. So we arrive at another where-to-draw-the-line question and once again, there is no simple answer.

The 9/11 complication The *Uniting and Strengthening America by Providing Appropriate Tools Required to Intercept and Obstruct Terrorism Act of 2001* (USA Patriot Act) was quickly designed in response to the coordinated attacks of September 11, 2001, in which four jet passenger planes were commandeered by a total of 19 terrorists from al-Qaeda, a militant Islamic group. Two planes were flown into the twin towers of the World Trade Center in New York City, bringing down both buildings. Another was crashed into the Pentagon building in Washington, DC. The fourth was forced down prematurely by passengers before it could reach its intended target in DC. All told, almost 3,000 people were killed and enormous damage was done to buildings, infrastructure, and associated services.

Signed into law by President George W. Bush just 45 days after the attacks, the Patriot Act greatly expanded the powers of law enforcement agencies to invade people's privacy by examining email, phone records, and financial information, by conducting wire taps, physically searching persons, homes, and businesses, and even probing lists of library books taken out, among many other provisions. Suspects, itself an ill-defined term, could be held without the full protection of due process. Immigrants could be detained indefinitely and deported based on suspicion of terrorism or association with known or suspected terrorists even if no actual acts were carried out or were being planned.

Many people opposed to the Act were concerned that in their zeal to increase security and protect the country from future attacks, Congress went overboard, unnecessarily weakening civil liberties. They pushed for reform, including making permanent the removal of several provisions that were due to expire at the end of 2005. Defenders of the Act pointed to its success as measured by the absence of new attacks and several potential attacks having been short-circuited. That was considered to be worth the price of reduced civil liberties.

The Act, in a slightly modified form, was reauthorized by Congress on March 2, 2006. About a week later it was again signed into law by President Bush. At the end of February 2010, President Barack Obama signed a one-year extension for some provisions that were due to expire. After several court cases, modifications were made by Congress to some of the other provisions of the Act to overcome claims of illegality.

A thorough examination of this quite lengthy and detailed Act[39] and the full range of reactions to it is beyond the scope of this book. Still, our brief foray into that discussion illustrates how measures to enhance security come at a cost to privacy and other civil liberties. The question is, what cost is appropriate? The US has a long history of coming down on the side of civil liberties when a tradeoff is considered. But placed in the context of the possibility of future 9/11-like devastation, which was of a magnitude not previously experienced in the US, the balance question is much more difficult to answer. Bringing cyber attacks into the picture adds another layer of difficulty.

Hard choices   The inescapable fact is that however we look at security we confront a tradeoff—the more secure we want to be, the more rights and freedoms we have to give up. Here is a simple way to look at it—the harder we make it for the bad guys, the more burdensome it is for the good guys. Striking a balance between being able to use digital systems as we wish and protecting us from digital attacks that can be incredibly disruptive is a fundamental issue that sits alongside protection from physical attacks.

Law enforcement authorities naturally are eager to use the newest IT tools to follow the movements of suspected miscreants. With GPS tracking, cell phone call and location records, E-ZPass data, and wireless monitoring equipment, patterns of movement, locations visited, trips taken, and purchases made, especially when compiled over time, can reveal all sorts of potentially incriminating and indictable activities, but also can make innocent activities look suspicious.

Because of the nature of the IT employed, digital surveillance, tracking, and deep data mining of collected information can be carried out surreptitiously, which, of course, is the point. But that raises a key issue: Does Fourth Amendment protection from "unreasonable searches and seizures" prohibit warrantless activity no matter what? After all, such classic enforcement agency methods as stakeouts, tailing, and phone call and card charge record searches can be warrantless. A number of questions arise: Does IT tracking constitute a search? Is data collection a kind of records seizure? Are the capabilities and effectiveness of digital surveillance and investigation so qualitatively different from traditional methods that proceeding without a warrant is constitutional? Is warrantless review and dissection of information unreasonable search?

The justification for these activities is that they are not used to cast a wide net to see who gets snared—only suspects are involved. But what defines a suspect? What leeway should authorities have to declare someone a suspect without judicial oversight? Regardless of how a suspect is identified, is warrantless investigation appropriate?

Purposeful tracking is not the only issue. Few are the places we can go without being under the eye of street cameras, office cameras, building cameras, bus and train cameras, red light cameras, license plate reading cameras, and more. Our online searches, purchases, viewing habits, travels, networks of friends and acquaintances, ailments and medications, leisure activities, and businesses we patronize—in short, everything about us—are recorded and kept indefinitely. Almost none of its capture or use is under our control and little of it is under the control of relevant legislation.

What do you think of this? *A narrow ruling*

The Fourth Amendment safeguards "The right of the people to be secure in their persons, houses, papers, and effects." Consequently, before the digital era, security focused mainly on physical intrusion of some sort. Digital IT makes it easy to get around that criterion, because searches can be carried out without physical trespass.

On January 22, 2012, the US Supreme Court answered one search question. The 5 to 4 decision concluded that placing a GPS tracking device on someone's car amounts to a physical occupation of private property for the purpose of obtaining information, so use of the device constitutes a search. Unless a warrant is obtained beforehand, that search is unreasonable, a violation of the Fourth Amendment.

The majority decision specifically engaged the physical requirement and did not address any other means of warrantless tracking, information collection, or search, nor did it address privacy issues.

Is electronic surveillance of any kind that doesn't involve physical trespass a way around Fourth Amendment protections? Must each issue be decided case by case? Can legislation governing situations that require warrants be better aligned with the capabilities of digital IT, making Fourth Amendment protection more inclusive?

We are in an era where digital attacks of all sorts can come from anywhere at any time. There is every indication that these kinds of threats will increase in variety, frequency, and capability. Consequently, countermeasures, both preemptive and after the fact, must engage the same kinds of increasingly proficient digital technologies. Among other things, that has to mean expanded surveillance and further loss of privacy.

Are we headed for a society where near total and constant surveillance is the accepted norm, where we all are fully known quantities? In the interests of security, do we have to be?

The Fourth Amendment

The right of the people to be secure in their persons, houses, papers, and effects, against unreasonable searches and seizures, shall not be violated, and no Warrants shall issue, but upon probable cause, supported by Oath or affirmation, and particularly describing the place to be searched, and the persons or things to be seized.

(http://caselaw.lp.findlaw.com/data/constitution/amendment04/)

Notes

1   Kahlil Gibran (1883–1931), philosopher: http://www.quotegarden.com/gossip.html
2   Chuck Palahniuk (1962–), *Invisible Monsters: A Novel*, WW Norton, 1999.
3   http://www.scmagazineuk.com/blaster-worm-remembered-six-years-after-it-hit-windows/article/146375/
4   http://news.cnet.com/Security-from-A-to-Z-Jaschan-Sven/2100–7349_3–6138562.html?tag=lia;rcol
5   http://www.pcworld.com/article/124472/botnet_hacker_pleads_guilty.html
6   http://www.scmagazineuk.com/russian-internet-blackmailers-jailed/article/106968/
7   A Trojan carries malware in disguise, as cargo. A fuller explanation appears in the Malware section below.
8   Dennis Blair, director of national intelligence, in a session before the Senate Select Intelligence Committee in February 2010, as reported in *Federal Computer Week*: http://fcw.com/articles/2010/02/02/web—dni-cyber-threat-annual-assessment.aspx
9   Reported by John Markoff, *New York Times*, March 28, 2009: http://www.nytimes.com/2009/03/29/technology/29spy.html?ref=johnmarkoff
10 http://www.msnbc.msn.com/id/43272064/ns/technology_and_science-security/t/senate-regularly-thwarts-cyber-attacks-official/
11 http://www.huffingtonpost.com/2011/06/01/google-gmail-hack-china_n_869995.html
12 http://www.huffingtonpost.com/2011/08/09/china-targeted-cyberattacks_n_921901.html
13 The study, sponsored by Juniper Networks, was carried out by the Ponemon Institute and was published in June 2011: http://www.juniper.net/us/en/local/pdf/additional-resources/ponemon-perceptions-network-security.pdf
14 http://www.whoishostingthis.com/blog/2010/08/25/first-draft-network-solutions-hacked-500000-domains-infiltrated/
15 http://www.citi.com/citi/press/2011/110610c.htm
16 http://host.madison.com/wsj/news/local/education/university/article_f5966aac-0408-11e0-af11-001cc4c03286.html
17 http://www.thelantern.com/campus/hacked-data-breach-costly-for-ohio-state-victims-of-compromised-info-1.1831311
18 http://www.credit.com/blog/2011/04/recent-govt-data-breaches-pose-privacy-risk/
19 http://www.credit.com/blog/2011/04/recent-govt-data-breaches-pose-privacy-risk/
20 http://www.nytimes.com/2011/09/09/us/09breach.html
21 A Web site where students can get or supply help with schoolwork, for a fee: http://studentoffortune.com/
22 http://online.wsj.com/article/SB10001424052702304772804575558484075236968.html
23 http://www.spamhaus.org/statistics/networks.lasso
24 http://krebsonsecurity.com/2010/09/spam-affilate-program-spamit-com-to-close/
25 http://krebsonsecurity.com/2010/10/spam-volume-dip-after-spamit-com-closure/
26 http://www.m86security.com/labs/spam_statistics.asp
27 For more information about DoS attacks, see http://www.tech-faq.com/denial-of-service-dos-attacks.html and for DDoS, see http://www.cisco.com/web/about/ac123/ac147/archived_issues/ipj_7–4/dos_attacks.html
28 http://www.pcworld.com/article/109174/nasa_servers_hacked.html
29 A flimflam is deception based on trickery, fast talking, and deals that seem to be too good to be true, aimed at swindling us into buying shoddy or fake products or betting on enticing but no-win propositions.
30 http://www.m86security.com/solutions/security_issues/web-based-malware.asp
31 For the full text of the CAN-SPAM Act, see: www.internetcases.com/library/statutes/can_spam_act.pdf
32 http://www.realuser.com/
33 http://www.captcha.net/
34 The ITU-T standard for PKI systems, including digital certificates, is X.509 http://www.ietf.org/rfc/rfc2459.txt
35 John Mariotti: http://www.goodreads.com/quotes/tag/%20malware
36 Richard Clarke, cybersecurity advisor to presidents Bill Clinton and George W. Bush, in an interview on NPR, April 2010: http://www.npr.org/templates/story/story.php?storyId=125578576&ps=rs
37 Benjamin Franklin: http://www.brainyquote.com/quotes/keywords/security.html
38 Bruce Schneier: http://thinkexist.com/quotes/bruce_schneier/
39 The full text of the Patriot Act is available through: http://www.gpo.gov/fdsys/pkg/PLAW-107publ56/html/PLAW-107publ56.htm

8

Mind and machine—artificial intelligence

Lunkwill: Do you . . .
Deep Thought: Have an answer for you? Yes. But you're not going to like it.
Fook: Please tell us. We must know!
Deep Thought: Okay. The answer to the ultimate question of life, the universe, and everything
   is . . . [wild cheers from audience, then silence]
Deep Thought: 42.[1]

Is it possible for machines to think, understand, exercise judgment, and learn as humans do—to exhibit human-like intelligence? Making that happen is the ultimate, though not the only goal of *artificial intelligence* (AI).

AI takes human intelligence as a reference point, so before we consider AI itself, let's look into the human variety. As we noted in Chapter 1, there is no generally accepted definition of human intelligence, but it is largely agreed that it is at least characterized by being able to:

- *understand*—interpret and make sense of what's going on
- *reason*—distinguish and select among interpretations or choices
- *learn*—integrate new information, improve understanding, and take reasoned appropriate action.

This triad is not the province of humans alone. Animals, insects, and other living things exhibit intelligence in that they understand their environments, learn to adapt to them, and make choices among actions, but not to the level that characterizes humans. That's considered to be our primacy.

Intelligence is not an all-or-nothing proposition. There are not only degrees of intelligence, but kinds as well, a wide range of types and levels among humans, animals, and machines too. Today's computer-based devices exhibit an array of intelligences, though they have yet to reach the point where machine behavior is indistinguishable from human behavior. Some say they never will.

In 1983, Howard Gardner popularized the notion of multiple intelligences with his list of seven types: linguistic; logical/mathematical; musical; spatial; bodily/kinetic; interpersonal; and intrapersonal. Later he added two more: naturalistic and existential.[2] Many education specialists have incorporated his ideas as they consider how people of all capabilities learn and how best to teach.

Of course, not everyone agrees with Gardner's ideas. Critics of his categorization say that he created them out of whole cloth and they have never been scientifically tested. Besides, they claim, what he is talking about is abilities, which are quite distinct from intelligence.

We could argue these and other points back and forth, but one thing is clear—defining human intelligence is not easy and must at least account for the different potentials and variations that characterize each of us. In the same way, then, when it comes to judging AI it doesn't make sense to believe that there's only one definition and one way to evaluate success or failure of machine intelligence.

It is also inconsistent that tests of AI tend to dismiss activities at which computers excel but humans don't, yet don't rule out any of the mental or associated physical activities of people. Sorting and mathematical calculations, for example, are easily done almost instantly by elementary computers but not by people. Computers do them quickly because they follow the steps of algorithms. Is that a reason to cross them off? If we think about it, people also follow algorithms when they sort and calculate—that's how we learned to make comparisons, to add, subtract, multiply and divide. Are not thinking and reasoning involved?

Tests of AI commonly focus on how successfully machines can carry out particular tasks and whether or not people can distinguish machine responses from human responses. Perhaps the most famous is the *Turing test*,[3] which its creator, Alan Turing, called "the thinking game." Three entities are involved: a computer, a human participant, and a human judge. The judge engages the other two in a natural language conversation and if he/she can't discern which is the computer and which is the human, then the computer is said to be exhibiting intelligence. (As first conceived, the conversation took place via keyboards and printouts so as not to be biased by computer voice recognition and speech.)

A little history

When AI began as a field worthy of serious research is not clear, but we can point to the Dartmouth Summer Research Project on Artificial Intelligence of 1956.[4] That spurred interest among many scientists and engineers, who started forming research teams and AI-focused institutes at universities.

The consensus is that the term artificial intelligence was coined by computer and cognitive scientist John McCarthy in his proposal for the Conference. He defined it as "the science and engineering of making intelligent machines." To more closely relate AI and human intelligence, at least for comparative purposes, let's say that AI is machines functioning with mental processes like those of human beings.

Since AI depends heavily on computers, it's safe to say that practical progress on the ideas coming out of the conference had to wait until computer capabilities reached a much greater level of performance than they had in 1956. And before any of the machine intelligence found in science fiction can be approached, another considerable leap in computing power will be needed.

AI is as many-dimensional and complex as human intelligence, so it was not possible to work on creating AI as a whole. Instead, different researchers focused on different areas, all more or less related to the grand scheme. Creating programming languages specially designed for intelligent tasks and improving hardware capabilities were early steps. Work in other areas also proved fruitful. Some examples:

- *Expert systems* build databases of the accumulated knowledge of experts in particular fields. People query the system and guided by its responses, make decisions in well-formulated situations that have many possible solutions —for example, diagnosing illnesses or ferreting out causes for equipment malfunctioning.

- *Heuristics* provide programmable logical ways of delving into problems that are not amenable to straightforward formulaic solutions. They are especially useful in situations where there are no concrete methods for finding an optimal solution save an exhaustive search of possibilities, which usually is not practical.

- *Machine vision* enables computers to take in and utilize visual information as a sighted person would, a requisite step in sight-based sensory evaluation. Computer vision has the advantage of not being restricted to visible light, as is ours. Computers can be designed to see infrared, ultraviolet, and other wavelengths of the electromagnetic spectrum that we can't. Intelligence, though, is more than simply sensing—it requires interpreting what is seen.

- *Fuzzy logic* comprises procedures for computer analysis of nebulous situations. When we are confronted with situations where there are no obvious ways to organize or formalize analyses and make decisions, we use our experience and familiarity with related matters to extrapolate from what we know to what might work. In a sense, we fill in the blanks. Fuzzy logic analysis combined with accumulated experience enhances the ability of machines to function well under uncertain circumstances.

- *Neural networks* imitate the biological circuitry and logic of the brain to capture the essence of the way humans process information for understanding, reasoning, and learning. They have been particularly successful in pattern recognition, especially when patterns are complex or embedded within other information. They are also proving useful for improving task performance by learning based on experience.

- *Mechanical/sensory-feedback systems* enable machines to perform actions like locating and picking up objects with sufficient grasping force but without damaging them, moving from place to place without bumping into anything, precisely placing them, and other sorts of physical manipulations.

- *Semantic understanding* facilitates computer comprehension of speech and writing so machines can respond in kind and act accordingly. That's a key component of human intelligence and therefore must be part of a true AI system.

- *Computer and storage technology* performance is increasing while size and cost are decreasing. AI systems that run on rooms filled with computers and storage farms are fine for developing the field, but to be generally useful, size and cost have to shrink dramatically as power grows.

In the years since the Dartmouth conference, some developments in AI showed promise but many, frustratingly, did not. As a result, the field has had many peaks and troughs in interest and popularity, at times dismissed as a wasteful pursuit of an unrealistic dream and at others as the inevitable path to the future. So where are we in the quest, over half a century later? There's no simple answer, and here's why.

By the 1980s, the experience of AI researchers solidified the realization that what we take for granted, even in young children—recognizing a picture of a friend, understanding the meaning of a cartoon, drawing and coloring—are very difficult for machines. At the same time, things that are easy for machines—alphabetizing huge lists, searching for specific words in a massive document, complex mathematical calculations—are easy for machines and very difficult for us. So, there is a great divide between tasks machines do well and tasks people do well.

That realization was captured by Hans Moravec, who writing in 1988 said:

> . . . computers are at their worst trying to do the things most natural to humans, such as seeing, hearing, manipulating objects, learning languages, and commonsense reasoning. This dichotomy—machines doing well things humans find hard, while doing poorly what is easy for us—is a giant clue to the problem of how to construct an intelligent machine.[5]

The explanation for what came to be called *Moravic's paradox* was a counterintuitive finding—what is called high level reasoning actually takes little computational effort, but low level mechanical/sensory manipulations need enormous amounts of computation. We process the great number of steps needed to pick a tomato out of a bin so quickly and naturally that we don't feel like we're thinking very much—which tomato looks the nicest (in our opinion), feels ripe but not too soft (also subjective), has a good size but isn't too big (ditto), how far to reach and position our hand to grasp it, how much pressure to exert with our fingers, how much force to lift it, how to hold a bag for the tomato, maneuver to drop it in, and put it in a cart in a spot where it won't get crushed by other items. That simple task calls for a lot of brain activity.

By contrast, sorting a column of 10,000 names seems to need a great deal of thought, but actually only requires making simple comparisons. The problem for people is not how to arrange alphabetically, but that there are so many comparisons to make.

The dichotomy was expressed by Steven Pinker who wrote in 1994,

> The main lesson of thirty-five years of AI research is that the hard problems are easy and the easy problems are hard. The mental abilities of a four-year-old that we take for granted—recognizing a face, lifting a pencil, walking across a room, answering a question—in fact solve some of the hardest engineering problems ever conceived.[6]

So there's the dilemma, and that's why sentiments about the future of AI have been encapsulated in the adage: "True AI is only twenty years away. The problem is, it's always twenty years away."

Whether or not that dictum ultimately proves to be true, there is no doubt that we are building increasingly intelligent machines, a fact that excites many people, frightens others, and moves some to states of denial. Joining debates about limits to machine and human intelligences are controversies revolving around ethical and moral issues, which take their place alongside the practical functional issues of AI.

A different view

The nearly two decades since Pinker wrote his assessment have seen exponentially great progress in computer power, machine sensory capabilities, and mechanical motion and manipulation skills. The faster–smaller–cheaper mantra of computers has not let up. Now it's entering the nanotechnology sphere. Meanwhile, machine sensory and mechanical abilities also are moving ahead. The field is expanding— intelligent computers, mobile intelligent devices, physically capable intelligent machines, all are in play.

Are there limits to these progressions? Whether or not there are, will we reach a point where what has so far been uniquely human intelligence and physical manipulation will be imbued in machines? What does it take to reach a practical level of human-like machine intellect and behavior?

The imitated mind   A computer, as a standalone intelligent device or as the brain of a quasi-human (robotic) machine, must be extremely powerful. Yes, it must process complex programmed instructions rapidly, but it also must be able to take in and evaluate enormous amounts of information, understand what it's doing, and learn from experience. Data storage must be huge, cross-classifications efficient, and retrieval swift. Whatever its size, for practicality's sake even the largest must be relatively inexpensive. To fit into a humanoid robot, it must be small—about human brain-size. These are tough requirements indeed.

What do you think of this? *Wetware vs. hardware*

In basic terms, the human brain, called wetware by technologists, is a biological organ comprising billions of neurons, each connected to others in intricate patterns by synapses. Those connections form electrical circuits that give us the ability to think, act, react, learn, remember, and all the other things we do. Some functions are autonomous—we don't have to think about breathing or making our heart beat—while others require our will.

Suppose we could construct a computer with the same billions of (artificial) neurons and synapses, give it some inborn capabilities (autonomous functions and instinct) by way of built-in circuits, and the ability to make its own connections as well—an artificial parallel to the human brain. Would there be any functional difference?

Sense and sensibility   Our five senses—sight, hearing, touch, smell, and taste—provide streams of information to our brains. For us to function, our brains must process those inflows and respond appropriately. That takes considerable brain power, without which inputs lie fallow.

Sight is a matter of light rays hitting the retina, but vision is much more than that—it must include understanding, making sense of those rays. Then they can be integrated into the task at hand—recognizing that a red light means stop and that a red brick flying at our head means duck. What we might call vision intelligence, a major component of general AI, already has played a role in cameras that stabilize images, automobiles that can navigate on their own, character recognition software, and image recognition and interpretation.

On the other hand the human eye–brain visual system can be confused by optical illusions. Parallel lines intersected by others look like they bulge. Intricate circular patterns appear to be moving though they are not. Colors shift according to how they're juxtaposed and displayed.

Some of the most remarkable illusions are specially constructed chalk drawings on sidewalks and streets. When viewed from the proper angle, their pseudo three-dimensional depictions are so incredibly realistic that they thoroughly confuse our visual sensory system into believing they are what they appear to be rather than what they are.[7]

Would computer visual systems see through illusions, to understand that they are not reality or should they be constructed to be fooled like us so as to be more human-like? These are not specious questions, because interpretation, a primary component of AI, is involved. Our brains may have a hard time understanding what we're seeing on first look at the sidewalk art, but we usually can draw simple conclusions from visual evidence that are not so easy to replicate in computers. For example, when we look at a piece of light-colored cloth with black polka dots, we know that the dots are part of the cloth and not holes in the fabric, and we usually can tell a stain on a patterned shirt from a part of its design.

What do you think of this? *Now you see it . . .*

Humans see patterns and images in all sorts of things: clouds, stains, fruit and vegetables, and even random squiggles. We also tend to anthropomorphize—attribute human or animal characteristics to non-living things. We use that tendency for purposeful representation—a smiley face sticker, a talking lemon head in a TV ad. We can see a face in a three-hole electrical outlet and animals in oddly shaped vegetables—and we can get carried away too. There are many instances of random patterns in burnt toast, beer foam, fried eggs, oil stains, and the like being seen as images of saints and more. Is that a characteristic we want to build into AI devices too?

From the true but hard to believe department: in November 2004, a piece of an old toasted cheese sandwich said to reveal an image of the Virgin Mary sold for $28,000 (http://news.bbc.co.uk/2/hi/americas/4034787.stm).

Interpretation for understanding applies to other senses as well. Hearing is more than sound waves causing our ear drums to vibrate—are we hearing music or a car horn? Touch is more than feeling something—is it the soft fur of a kitten or the sharp edge of a knife? And are we smelling a delicious peach or a rotten egg? Does that morsel in our mouth warrant swallowing or spitting out?

Our senses work with our brains in feedback loops. If we want to pick up an object, we judge how far to reach and once we get to it, how hard to grasp. The force required to grip a ten-pound weight would crush a banana, but as we see each one and begin to lift it, continuous feedback helps us make the right moves.

Our brains also work to fill in sensory gaps and clarify inputs. We can identify a blurry picture of a friend even though it's not immediately apparent. We envision where a jigsaw puzzle piece might fit. We know that a siren going off at a football game to celebrate a touchdown is not the same as a siren signaling a fire in a building. The breeze in our face carrying aromas of a bakery evokes images of breads and pastries. We understand that a phone ringing in a TV show is not for us to answer.

Computer brains need the same sorts of sensory input and feedback to their processing systems, plus interpretive capabilities to understand what's going on. Of course, not every intelligent machine needs every sensory or interpretive capacity, but we can imagine machines that have them all. Moreover, there's no reason why the abilities of machine sensory systems can't exceed ours.

We also can envision machines whose manipulative abilities surpass ours. These are particularly difficult to create because many of our movements are inherently intricate and involve a great deal of co-ordination, needing not only mechanical skill but a high level of brainpower.

Progress on many fronts Work on single senses and skills is ongoing, along with efforts to integrate them into more competent machines. The mechanisms needed for the devices to do their work are improving, as are the "bodies" that will contain everything.

Nature, having solved a lot of the problems of motion and maneuvering, makes living creatures of all sorts good models for building those abilities in robotic devices. Called biomimicry, some promising work is taking place.

Research into the way animal and human brains work provides valuable insights for making progress in computer intelligence. That is leading to machines that are controlled by artificial

> **Working definitions**
>
> *Automaton*: a general term for a complex self-contained machine that can sense and react to its surroundings, and exhibit various levels of intelligence and mechanical capabilities.
>
> *Robot*: a kind of automaton. Industrial robots, usually running software designed to perform particular tasks, rarely have any resemblance to humans; other robots have various human-like characteristics.
>
> *Android*: a highly intelligent automaton, most often a robot with at least a passing resemblance to humans and human-like capabilities. Advanced androids would be quite human-like in appearance, brain power, and functional capabilities.
>
> *Cyborg*: an automaton composed of artificial and biological components, a melding of human and machine with capabilities exceeding those of more traditional robots or androids, and humans.

human-like guidance systems and those that operate via human–machine interfaces, computers with semantic understanding and those that learn.

Consider these examples of directions and progress:

- *Robotic animals and insects* mimic physical movements of living creatures in mechanical devices. Some success has been had with fish, snakes, salamanders, and dogs, among others, all of which must be able to sense and react to their environments as they move the way the creatures do.[8]

  An interesting offshoot is *robotic pets*—usually cats, dogs, rabbits, and other small furry versions of animals—that respond to being held and touched. The idea is to evoke calming and warm emotions in people to help them cope with their predicaments.[9] Anecdotal evidence is promising.

- *Rubot II* is a robot that was specifically designed to manipulate and solve physical Rubik's cubes. Solutions are algorithmic, but solving the cube requires being able to see and recognize the patterns of the cube's colored blocks and swivel them accordingly.[10]

- The University of Tokyo's Ishikawa Oku Laboratory has developed amazing mechanical devices, one of which is a *three-fingered hand and arm* that can twirl a pen, tie a knot in a string, and dribble a small ball, all so rapidly that it's hard to follow. It also can throw a ball to hit a target, use tweezers to pick up and deposit a grain of rice without hesitation, toss a cell phone into the air and catch it on its fingertips, and more.

  Its incredible quickness, dexterity, and other abilities come from computer-coordinated high speed mechanics, integrated vision, tactile sensitivity to properly grasp and control different kinds of objects, and what in humans is called hand–eye coordination.[11]

> The word robot first appeared in Karel Capek's 1920 play *R.U.R* (*Rossum's Universal Robots*, or *Rossumovi univerzální roboti* in the original Czech). It comes from robota, translated as a worker or serf. These robots were biological creatures hard to distinguish from humans, who did the work while humans lived at leisure.
>
> Eventually the robots took over, killing all humans but one. However, the robots were headed towards extinction when the formula for creating them was lost. The surviving human strove to recreate the formula, but in the end, relied on two young robots, male and female, to become a new world robotic Adam and Eve.

- Boston Dynamics' *Big Dog* is a four-legged 240-pound mule-size robot that can navigate all sorts of terrain, level and hilly, sandy, rocky, ice and snow covered. The Dog can jump over obstacles, maintain balance, recover from slips and buffeting, and do all that while carrying a load 100 pounds greater than its own weight. It's controlled by arrays of sensors feeding an onboard computer whose programs include feedback loops, enabling it to react rapidly to changing conditions.[12] Boston Dynamics also is developing a two-legged version called *PETMAN*.[13]

- Aldebaran Robotics' *NAO* is a two-foot high two-legged two-armed robot that can maneuver in various ways, none of which will be mistaken for human movements, although in their ungainly way they accomplish some of the same things. It has tactile sensing fingers, vision, hearing, voice response, and balance capabilities, integrated via its onboard computer. Importantly, it can communicate with other NAOs.

Unlike the somewhat scary looking Big Dog, NAO is undeniably cute, given pseudo human features like a swiveling head with two large eyes and big earlike openings. Though the "eyes" are cameras, the "ears" sense sound, and the head swivels to broaden its field of vision, none of those capabilities needs to appear human-like to function. But NAO's endearing appearance makes it easy to accept and even relate to it.[14] Some robotics developers believe that is important, as people are unlikely to feel threatened by charming machines.

- *Asimo*, a humanoid robot build by Honda, is the most capable of the android types. Also built with two legs, two arms, and a head, it can walk forwards, sideways, and backwards, alone or hand in hand with a person, dance in time to music, run in straight lines and around curves, walk up and down stairs, follow corridors, avoid stationary and moving obstacles, carry and deliver items, coordinate with other Asimos, and interact with people in various ways. Perhaps the most impressive of Asimo's skills is its ability to learn and generalize.[15]

  Asimo, about twice the size of NAO, is another of the cute robots. Though clearly not human, it not only has an engaging look but also has some endearing actions. For example, if you step close to Asimo and call its name, it will reach out in a sort of friendly welcoming gesture. That kind of emotion-seeming interaction is a large part of Asimo's appeal, just as was true of R2D2's and C3PO's.[16]

Emotional expression, whether by people or computers, and whether for understanding or responding, is another avenue of research.

- *eXaudios Technologies* has found that voice intonations contain clues to people's moods and emotions. They have developed computer programs that can determine a speaker's moods and emotions from as little as 30 seconds of speech, without any visual clues. The programs "analyze voice and determine . . . the instinctive-emotional personality of the speaker and his emotional attitude." The analysis is "based on established links between emotional attitudes and physically measurable parameters in voice." [17]

  eXaudios' primary goal in developing the programs was to use them for improving marketing and sales by understanding the underlying emotional states of prospective customers. However, while analyzing thousands of voice streams, they found that people with such conditions as dyslexia, Parkinson's, and autism had distinctive markers in their voices, leading to the possibility of using the programs as intelligent computer-based diagnostic tools. Joint projects with medical researchers are underway.

- The *Affective Computing* group at MIT explores a variety of aspects concerning emotional expression. Among them are "machine recognition and modeling of human emotional expression, machine learning of human preferences as communicated by user affect, intelligent computer handling of human emotions."[18]

Their overall aim is "to bridge the gap between human emotions and computational technology." That can go a long way towards building machines that understand the context of human communication, just as we do when we know the emotional state of a person we're speaking with. It might also help machines and people relate to each other in a more natural way.

Facial expressions telegraph emotions and can give away suppressed thoughts. Can our ability to recognize and interpret them become part of the AI picture? Developments are proceeding along two lines.

Visual input combined with pattern recognition software and facial expression databases can give computers the ability to distinguish and interpret facial expressions. That has applications as wide-ranging as computer identification of criminals from surveillance videos, detecting lies during interrogations, having your car recognizing you and starting without a key or fob, having your computer booting up without your intervening or needing a password. The next step is for computers to add to their store of expressions, learning from experience as we do.

- Researchers at the Universidad Politécnica de Madrid and Universidad Rey Juan Carlos have created algorithms that can recognize facial expressions fast enough to operate in real time, as do people. So far, computers running their software can interpret emotions of anger, disgust, fear, happiness, sadness, and surprise.[19]

From the opposite perspective, realistically human-looking robots are being built that can mimic facial expressions, eye movements, and gestures, speak and respond to questions. Their personae, perceived as non-threatening, approachable, and even friendly, are believed to be important for people to accept android robots.

That view is bolstered by the reactions observed when people have seen these sorts of robots at trade shows and conferences. But it's not universal. Some feel that such realistic androids are creepy and prefer the cute but clearly non-human looking robots if they have to deal with them at all.

Progress is slow and extremely difficult because of the complexity of the human facial structure and musculature, but progress is being made. Early uses for these kinds of robots are as greeters, ticket sellers, information kiosk workers, and the like. As they become more capable, they are being considered as aids to health care workers, interacting with patients.

- *Repliee Q1*, produced by the Intelligent Robotics Lab in the Graduate School of Engineering of Osaka University, has a strong human resemblance, basic facial expressions and hand/arm movements, and the ability to speak and respond to questions. Developments in tiny actuation systems, flexible structural components, artificial skin and eyes, and semantic understanding will lead to quite realistic androids.[20]
- Geminoid robots look eerily like clones of their creators. One example is *Geminoid-DK*, developed in a project led by Professor Henrik Scharfe, Aalborg

University, Denmark. The android, which has a remarkable resemblance to the professor, is designed to mimic his motions as he does them. At this juncture, movements are quite basic, but are bound to improve with further work in what is another offshoot of AI development—interaction between people and machines.[21]

Of course, not all AI-imbued devices have to resemble anything living. At present, those that don't are in the overwhelming majority. Some, like smartphones, simply have to be functionally effective and easy to use. Others are meant to operate behind the scenes in special facilities, on production floors, and embedded in equipment and appliances.

Brain-machine interfacing is another pursuit. Experiments in linking human brains with computers and computer-controlled equipment is one avenue of research. Another is using living brain cells implanted in mechanisms to aid in machine awareness. These are two of the paths being explored to understand brain functioning, create more capable computer brains, and show pathways to human/computer interfaces. Eventually, people will be able to issue commands to computer-controlled devices just by thinking about their desired actions. Some progress has already been made.

- Reading University, UK, has incorporated rat brain cells into a small robotic device that help it react to obstacles based on input from the device's sensors.[22] Researchers at the universities of Michigan and Arizona have worked on "cyborg insects," living insects whose movements can be computer-controlled via impulses sent to electrodes attached to their brains.[23] These experiments are first steps in human/computer symbiosis.
- A research group at the University of Pittsburgh led by Dr. Andrew Schwartz has taken another step. The group has taught a monkey to control a robotic arm using only his thoughts. Implanted electrodes transfer the monkey's brain signals to a computer that turns them into actions of the arm. Professor John Donoghue and his group at Braingate Research have applied the same concepts to enable a paralyzed man to control a computer with his thoughts.[24]
- A team at École Polytechnique Fédérale de Lausanne led by Professor José del R. Millán has developed a brain–machine interface that a person can use to control a wheelchair by thinking about directing it.[25]
- Similar developments were reported in a CBS *60 Minutes* episode in which brain waves from thinking about individual letters were relayed to a computer, enabling an immobile person to write.[26] Another program piece shows early steps in peering into the brain using functional MRI.[27] The patterns displayed when persons thought about various objects were able to be interpreted by a computer, revealing which objects the person was thinking about.[28]

These advances illustrate two avenues of brain/machine interfaces—one in which a computer is interposed between a person's thoughts and their physical representation, the other whereby a person's thinking itself is exhibited. Intelligent systems using these interfaces can act in accordance with a person's thoughts, or disclose them.

Both abilities can be viewed as forms of mind reading—sensing and interpreting the electromagnetic emanations or metabolic activity of brains as they think. That could lead to a more robust understanding of how a human brain's circuits produce thoughts and control actions, which then could be copied in designing computer brains.

What do you think of this? *Planting ideas*

Suppose mind reading reaches a stage where sensors placed surreptitiously could divine your brain waves as you pass nearby and send that data to AI computers that would use your thoughts to influence your behavior. That might improve the effectiveness of targeted ad messages to increase the likelihood of your making a particular purchase or to vote for a certain candidate. The same technology also might be used to help patients overcome anxieties, addictions, or inappropriate behaviors. Should research and development on this type of IT be pursued unimpeded or should a line be drawn between what are acceptable avenues of research and what are not?

A quite different branch of the AI journey is the design and construction of *autonomous vehicles*. Research and development initiated by DARPA[29] spurred interest in the creation of driverless vehicles that could navigate on their own to reach particular destinations. Success requires a variety of sensory systems, input processing, and machine-directed behavior based on real-time analysis—machine intelligence.

To encourage the work, DARPA established the *Grand Challenge*, an event in which teams who built such vehicles could compete for prize money, showcase their skills, and possibly garner military contracts. Designs and vehicle types were unrestricted as long as they ran autonomously. So-called drive by wire remote control, the way drone airplanes operate, was not allowed.

- The initial competition, held in 2004, had a $1 million prize and required the entries to run a 150-mile course through the Mojave desert. None of the more than 100 vehicles completed the course. In fact, the prize winner was a team from Carnegie Mellon University whose vehicle managed fewer than 7.5 miles, but that was more than any other.
- For the second event, held in 2005, first prize was increased to $2 million and the course was changed to a series of roads, mountain passes, narrow tunnels, and sharp turns, with various obstacles along the way. Entries ranged from a motorcycle to a 30,000-pound truck. After eliminations, 22 of the 23 finalists finished the course, including the truck, which barely fit on some of the roads.

  The winning team, from Stanford University, ran the course in just over 6.5 hours; Carnegie Mellon had two entries, one of which came in second, only 11 minutes behind; the other took third, 9 minutes later. The truck, which had some very careful maneuvering to do, finished in about 12.5 hours.

- The last Challenge was held in 2007. Prize money for second and third places, $1 million and $500 thousand, was added to the $2 million first prize. The course was changed again. This time the vehicles had to travel 60 miles over regular streets, navigate the route appropriately, obey all traffic laws, be aware of other vehicles on the roads at the same time, cross intersections, and park. Carnegie Mellon's team won, finishing in just over 4 hours and 10 minutes. Stanford, taking about 20 minutes longer, came in second, and Virginia Tech, seven minutes behind, was third. Several more teams also completed the course.

Only three years after the first competition, tremendous strides had been made. The vehicles had to sense and comprehend their surroundings, decide where, how, and when to turn, when to slow down, speed up, and stop, how to read and understand traffic control signs, signals, road markings and curbs, follow lanes, interact with other vehicles, and more—in short, everything that people do when they drive except be distracted by cell phones, texting, and the like. Clearly, considerable intelligence was required—not full human intelligence, but impressive nonetheless.

- *Google* embarked on a driverless car project of its own, which is still ongoing. They enlisted Sebastian Thrun, whose Stanford University team won the 2005 Challenge, to lead the work. Their vehicles run over regular roads amidst regular traffic and do quite well. Only one was involved in an accident, and that's because it was rear-ended, no fault of the car's. The combination of Google influence and vehicle performance led Nevada to become the first state to permit the legal operation of autonomous vehicles on their roads.[30]

Semantics, the ability to grasp the meaning and intent of natural language discourse, is a normal part of human development that begins in infancy. It also is a major hurdle in creating machine intelligence and a foundation piece for what is being called the semantic Web—one that understands our queries intuitively and responds in kind—presumably the next phase in its evolution.

That kind of semantic processing goes far beyond the voice response systems we encounter when we phone customer contact centers or help lines, which by comparison are quite elementary. They only work when specific words are used in response to system queries, so they don't really understand—they aren't semantic interpreters.

- The most advanced semantic processing computer system so far is IBM's *Watson*. In operation, Watson can understand questions spoken in natural language and respond in kind. Behind that simple sentence is a huge amount of work, some very clever programming, a vast amount of data, and powerful computers.

  Four years in development, Watson's achievements are a remarkable leap ahead and a major breakthrough in AI. Here's why. Watson is not a look-up-the-answer system and is not connected to the Internet. It is an intelligent machine that operates semantically much the way humans do. It interprets even ambiguous questions, organizes possibilities, determines likelihoods, and responds with a specific answer to each question which, if not confident of, it acknowledges may be wrong.

Questions can include subtle clues, puns and ironies with multiple meanings, everyday slang, and other indirect references—things that people would understand but no other computer system would. Watson's thinking processes are akin to those of humans, quite different from search engines that supply lists of results that can range from relevant to totally off the mark.

A major test for Watson was its participation as a contestant on the TV quiz show *Jeopardy* in January 2011 (aired in February). Pitted against two human opponents, one of whom was the biggest money winner in the show's history and the other who had the longest winning streak, Watson won handily.

More remarkably, Watson ably decided when to take a chance on an answer that might be wrong, and determined how much of its winnings to risk on "bonus squares" and "final questions" based on question categories and its own and opponents' winnings—in other words, Watson was strategizing.

Watson is now being tested for use in aiding medical diagnoses. As it improves, more fields will be explored.[31]

Becoming ordinary

We are surrounded by AI of all sorts right now—low level compared to full-blown AI, but AI nevertheless.

- Dishwashers and washing machines sense how large and how soiled their loads are and adjust their cycles accordingly.
- Improved search engines make intelligent decisions about what to display and how to find their way through vast numbers of possibilities.
- Warehouse systems manage inventory by tracking usage, automating storage, retrieval, dispensing, and reordering, some utilizing robotic machines that handle the items from receipt to order fulfillment and distribution.
- Electrical grid monitoring systems shift electrical supply within and among interconnected grids in accordance with varying loads and generating capabilities.
- Airplanes take off, fly, navigate, and land using autopilots.
- Satellites are guided into rendezvous with various orbiters, many orders of magnitude more difficult than maneuvering earth-bound vehicles.
- Weather forecasters rely heavily on smart computer models.
- Traffic signals adjust automatically according to traffic flows.
- Telephone calls and Internet traffic are intelligently routed through complex networks as loads fluctuate.
- People working on projects as diverse as developing intricate packaging, architecting buildings, and designing computer chips are aided by intelligent programs.

● Products from airliners and automobiles to circuit boards and computer chips are manufactured and assembled in great part by computer-guided equipment.
● Smart data analysis and pattern recognition aid in catching criminals, diagnosing diseases, tracking potential pandemics, interpreting medical scans, validating passport applications, catching computer malware, and sending targeted ads.

Every day, computer intelligence becomes a little more capable, finds a few more applications, and spreads its influence a bit farther. Just as we rarely think about how AI is incorporated in so much of today's world, so will we take as given advances that provide all sorts of services but actually operate largely unseen in the background.

In the same way that our dependency on electricity goes unnoticed until a power outage brings it to our attention, so will the services provided by intelligent machines as they become ordinary utilities—we'll expect them to be there but will be aware of our reliance on them only when they don't work.

And so—master or slave?

> Knowing a great deal is not the same as being smart; intelligence is not information alone but also judgment, the manner in which information is collected and used.[32]

There is a great reluctance on the part of many to ascribe intelligent behavior to machines even when they do exercise judgment, reasoning, and learning. Perhaps it's due to a fear that computer-based machines will someday control and supplant us at the top of the evolutionary chain, or because of an unwillingness to believe that any device, especially a non-biological one, however capable, can truly exhibit something as uniquely human as higher order intelligence.

Here is another way to think about it. Biologically speaking, intelligence, consciousness, awareness, and the ability to debate and pose questions about them are products of electricity flowing in the circuits formed in our brains. That the biological basis of our brains is the differentiator between human and machine intelligence is questionable. If brain circuitry, both pre-established as in newborns and self-created as learning takes place, can be duplicated in silicon or whatever other substance, then logically speaking intelligence, consciousness, and awareness will be brought along with it.

Suppose a biological foundation does turn out to be a key distinction. Scientists can now create elementary living organisms. As that field advances, artificial brains could be composed partly or totally of biological components. Would that resolve the dispute?

Research and development activity in AI is on the rise. Advances unthinkable just a decade ago have been achieved. There is a long way to go in terms of true human-like machine intelligence, yet even now as AI grows more powerful, we face another question of limits: Where should a line be drawn between what is permissible development activity and what is not, or should there be no line?

To establish meaningful boundaries we need to know what to expect from a line of research. But AI, perhaps more than any other technology, has the potential for important possibly game-changing unintended consequences, positive and negative. Drawing any line, even for research that at first blush seems to be laden with ethical issues, might forestall major positive benefits.

Other issues arise. Is there a difference between a brain and a mind? What about a conscience? What about a soul? Those questions are more appropriate for philosophers than scientists and engineers. When the day comes that interactions with and among intelligent machines are indistinguishable from interactions with and among humans, those questions and the possibility of higher order AI will have been answered.

Computing power—the end of the line?   No matter how we view the prospects for true AI, we know that approaching it will take a big jump in computer capabilities. A common measure of computing power is the number of transistors that can be stuffed into an integrated circuit. For over 50 years, that number has doubled about every 18 to 24 months, a rule of thumb that has come to be called *Moore's Law*.

That has been possible because of continually shrinking the size of silicon-based transistors and refining chip production methods. The result, an enormous surge in computing power and concomitant drop in size and prices, has fueled the digital age. There is some question as to whether that progression, vital to the realization of true AI, can continue.

One issue is that the laws of physics will dictate an end to Moore's Law for the kinds of computers we're building today. Once we've reached the limits of shrinking silicon, we can go no farther, a point we might reach in as few as fifteen years. That being the case, something else needs to take over if we are to proceed.

Possibilities include: nanotechnology, working with molecules; sub-nanotechnology, working with atoms; and quantum computing, working with invisible units of energy called quanta. While those are hypothetically feasible, there likely will be a hiatus of many years before computers based on their theories can be designed and built. But if that day comes, we can resolve Moravic's paradox.

Day to day   The IT underpinnings of societies are increasing in complexity and they are bringing us along for the ride. We've moved from a time when most interactions were neighborhood based to the interconnected world of today, and from when technology was mostly in the hands of a relatively few engineers and scientists to being at the fingertips of all sorts of us.

Now, everyday living is intertwined with AI. As developments inexorably continue, IT and AI will become more multifaceted and we will become increasingly dependent on both, no matter our walk in life. Without the help of AI we either will falter or miss out. Just think of the impossibility of handling the growing flood of information minus the aid of intelligent processors. If we must labor under a deficit in AI, our growing abundance of data resources will force a decline in the scope of our understanding.

So we need to avoid being inundated with information, but how? Do we build dikes to block the flows or utilize computer intelligence to do some guided filtering?

That's a question that applies to people engaged in every profession and to our personal lives as well, to a broad spectrum of income levels and education levels, from the most isolated among us to the most extroverted.

This is an issue of huge proportions, because it relates not only to information flow but to almost everything we're involved with. We need AI to help run utility infrastructures, transportation systems, businesses of all sorts, financial systems, health care delivery and practice, manufacturing, provision of services—the list is endless.

So unless we go back to the pre-digital world, we have no choice but to become increasingly dependent on AI, which is bound to be embedded in more of the technology we use. Can we expect that AI will give us more control over the inter-woven systems of technology we face or will we have to trade away some significant amount of autonomy as technological complexities become too much for us to handle alone? Will we be in control, or maybe more ominously, will AI take over?

Mind and matter

Artificial Intelligence[is] the art of making computers that behave like the ones in movies.[33]

Any sufficiently advanced technology is indistinguishable from magic.[34]

Machines that can access, manipulate, create, utilize, and dispense information can be viewed as boon or doom. When we talk about the potential of AI to reach and especially to supersede the capabilities of human intelligence, opinions become polarized, sentiments grow more emotional and intense.

Reasoned logical scenarios are being posed where, in the not too distant future, machine intelligence will exceed that of humans. This does not simply mean computers that can come up with factual answers to a panoply of questions in the blink of an eye. Instead, what's foreseen is a time when machines learn from their mistakes and experiences, adapt to changing conditions, are comprehending, perceptive, have emotions, and even reproduce—in other words, human-like or superhuman-like. That puts us in a whole new ballgame.

The possibilities are incredible, ranging from computers smart enough to understand us and assist with our endeavors to computers and humans blending into cyborgs with extra-human capabilities. For now, this sounds like the stuff of science fiction, of movies replete with special effects showing epic struggles between machines and people, of pipe dreams and pipe nightmares. But consider that over and over, the science fiction of yesterday is the technology of today.

> The two most common tracks in sci fi stories about intelligent creations are:
>
> - The creator and the creation are destroyed when the creation goes bad and either destroys its creator in an epic battle before succumbing, or the creator wins, but must pay for arrogance in "playing god" with his/her own life.
> - Good and bad creations battle each other, with the good ones eking out a victory, usually with the help of humans.
>
> Let's hope neither of these scenarios is where we're headed.

So, we can envision a world wonderful or nightmarish. Who or what will be in control? Will advanced AI machines serve us or will we serve them?

Fears have led to thoughts of limiting AI development, the same sort of fears that have been used as arguments to keep other technological advances in their places or stopped altogether.

Some believe that we must build a cutoff switch into every artificially intelligent creation, a limiter that prevents machines from going beyond a certain level of competence and will shut them down if they go too far. Others say the march ahead will bring unimagined benefits; rather than being impeded, research should be vigorously supported.

What do you think?

Notes

1   From a conversation with the super computer, Deep Thought, specially designed to answer the question, in Douglas Adams, *The Hitchhiker's Guide to the Galaxy*, Pan Books, 1979.

2   Howard Gardner (1943–): http://www.infed.org/thinkers/gardner.htm

3   Alan Turing (1912–1954), "Computing machinery and intelligence," *Mind*, 59 (1950): 433–460.

4   Organized by John McCarthy (1927–), Marvin Minsky (1927–), Nathaniel Rochester III (1919–2001), and Claude Shannon (1916–2001): http://www-formal.stanford.edu/jmc/history/dartmouth/dartmouth.html

5   Hans Moravec, *Mind Children: The Future of Human and Robot Intelligence*, Harvard University Press, 1988, p. 9.

6   Steven Pinker, *The Language Instinct: How the Mind Creates Language*, William Morrow, 1994.

7   For a sample of several, see: http://www.thedailyweird.com/category/sidewalk-chalk-art/

8   See videos of: a robotic snake: http://www.youtube.com/watch?v=cJuNe50uuzk&feature=related; a robotic dog: http://www.youtube.com/watch?NR=1&v=B0qYob_vSgo; and a variety of devices: http://www.youtube.com/watch?v=Tq8Yw19bn7Q

9   Examples of "mental commit" robots are at: http://www.youtube.com/watch?v=AK-2laSYpII

10   For a video of Rubot II in action, see: http://www.youtube.com/watch?v=jkft2qaKv_o

11   For a video of the hand in action, see: http://www.youtube.com/watch?v=-KxjVlaLBmk For additional videos, see the Lab's site at: http://www.k2.t.u-tokyo.ac.jp/movies/fusion_movies-e.html

12   For videos of Big Dog in action, see: http://www.youtube.com/watch?v=Vu2IXk5Jbag; and http://www.youtube.com/watch?v=W1czBcnX1Ww&feature=related

13   See: http://www.youtube.com/watch?v=YX93nDOupJU&feature=related

14   For a video of NAO in action, see http://www.youtube.com/watch?v=rSKRgasUEko

15   Videos of Asimo: giving right of way: http://www.youtube.com/watch?v=yJf8Z6k4eyk&feature=related; avoiding another Asimo: http://www.youtube.com/watch?v=OSp1R6uanfQ&feature=related; avoiding stationary and moving targets: http://www.youtube.com/watch?v=YPoANTKo5kA; interacting with another Asimo: http://www.youtube.com/watch?v=w7G_pyojs5I&feature=related; and learning: http://www.youtube.com/watch?v=P9ByGQGiVMg

16   Intelligent robots appearing in the movie *Star Wars*, written and directed by George Lucas, Lucasfilm and Twentieth Century Fox Film, 1977.

17   http://exaudios.com/

18   http://affect.media.mit.edu/ http://affect.media.mit.edu/

19   Reported in: http://www.sciencedaily.com/releases/2008/02/080223125318.htm

20   See Repliee in action at: http://www.youtube.com/watch?v=09UQlGD-qbw

21   A video of Geminoid-DK narrated by Scharfe is at: http://www.youtube.com/watch?v=NSLe7xrP4jQ

22   See a short video at: http://www.youtube.com/watch?v=rSKRgasUEko

23   A short video is at: http://www.youtube.com/watch?v=dSCLBG9KeX4&feature=channel

24   For an explanatory video of both projects, see: http://www.youtube.com/watch?v=7kctOHnrvuM&feature=related

25   An explanatory video is at: http://www.youtube.com/watch?v=0-1sdtnuqcE

26   A video narrated by Scott Pelley is at: http://www.cbsnews.com/stories/2008/10/31/60minutes/main4560940.shtml?tag=currentVideoInfo;videoMetaInfo

27 Functional Magnetic Resonance Imaging (fMRI) uses high powered magnets to scan the brain, producing maps of brain activity. The maps are images that highlight the areas of the brain that are active when particular tasks are performed.

28 A video narrated by Leslie Stahl is at: http://www.cbsnews.com/video/watch/?id=5119805n&tag=cbsnews MainColumnArea.5

29 DARPA, the Defense Advanced Research Projects Agency of the US Department of Defense, was established to fund research in a variety of areas: http://www.darpa.mil/ (and see also Chapter 3).

30 See: http://www.youtube.com/watch?v=eXeUu_Y6WOw; and http://gizmodo.com/5662005/test-driving-googles-driverless-car

31 http://www-03.ibm.com/innovation/us/watson/what-is-watson/index.html

32 Carl Sagan (1934–1996), astronomer and astrophysicist: http://thinkexist.com/quotation/knowing_a_great_deal_is_not_the_same_as_being/198021.html

33 Bill Bulko: http://www.quotes.net/quote/11415

34 Arthur C. Clarke, "Hazards of Prophecy: The Failure of Imagination," *Profiles of the Future*, Warner Books, 1985. (Originally published by Harper & Row, 1962.)

9

We are what we do—ethics and responsibilities

In law a man is guilty when he violates the rights of others. In ethics he is guilty if he only thinks of doing so.[1]

Even the most rational approach to ethics is defenseless if there isn't the will to do what is right.[2]

Ethics, easy to invoke and hard to pin down, have been treated over a long period of time by a diverse agglomeration: philosophers, politicians, business associations, professional organizations, religious leaders, consumer groups, and each of us in our roles as members of societies and people engaged in all sorts of activities. In the preceding chapters, we've seen that technology in general and information technology in particular often bear ethical issues, especially when it comes to drawing a line between what can be done and what should be done.

It would seem that ethical behavior is the solution to setting boundaries. When confronted with a choice, simply do the ethical thing, the right thing. Not the least of the issues is that "should" and "right" are freighted, highly subjective notions. Ethics itself is not a universal independent concept either. Then too, choices may conflict on ethical dimensions, making more than one possibility simultaneously ethical and not.

Some ethical questions are easy to resolve and others, even seemingly straight-forward ones, are difficult. Nevertheless, we need to consider the role that ethics plays in our choices. That doesn't mean we'll always come to the right conclusion, if there is such a thing. Even if there is, we may not have the information we need to choose correctly, especially when the effects of a choice may not be known for some time. Still, if we are to advance, individually and societally, ethical dimensions must play a part.

A little background

Philosophers have grappled with ethics since the beginning of philosophical discourse and have come up with many ways to look at ethical behavior. At base, they all have the concept of right and wrong, of doing what's proper in a given situation. That's easy to say but not always easy to understand. As you might expect, there are almost as many avenues of thought as there are philosophers, but they can be condensed into three broad areas: metaphysical, normative, and applied.

- *Metaphysical ethics* seeks what are believed to be fundamental truths, the foundations of ethical principles, hence the basis for ethical behavior. The idea is to clarify the concept of ethics by unveiling its sources, hence its very nature. Issues revolve around the meaning of good and bad, right and wrong, moral and immoral behavior, and providing fundamental support for making judgments.
- *Normative ethics* is concerned with what should be done, establishing rules for behavior and standards of values that serve as guides when faced with making ethical judgments or decisions with ethical consequences. It is less concerned with meaning, the focus of metaphysical ethics, than with what to do in various circumstances. So, it attempts to provide principles that can be models for appropriate ethical behavior.
- *Applied ethics* is where the rubber meets the road. It looks at how to use ethical guidelines and behavioral standards in specific situations as a means of comparing options and outcomes, individual and societal effects, and thus making proper choices.

Taking off from the philosophical foundations of ethics, we find morality, integrity, and lawfulness. All of these circle around the same ideas, adding a twist or shading.

- *Morality* requires following a prescribed code of culturally based behavior. Such codes often are grounded in religious views, although religious precepts themselves are not requisite to acting in a moral manner. Religions tend to look at their moral philosophies as being universal, which leaves little room for societal, religious, and cultural differences.
- *Integrity* means consistently following an ethical code of behavior. It assumes that the code is appropriate in societal terms and usually includes being honest, sincere, and not making exceptions just because it's convenient to do so.
- *Lawfulness* simply means acting in accordance with the laws of the land. That in itself can cause conflicts when laws contradict individual ethical or moral codes.

Some major concepts   As would be expected, philosophers and ethicists have different ideas about classifying ethics and ethical behavior. The following is a representative sample. The brief summaries are meant to express their main ideas, not the nuances and complexities of each concept.

- *Cultural relativism*—do what meets social standards, what is considered to be socially accepted behavior. Specific to individual societies, strict adherence to cultural relativism leaves little room for dissent or disagreement.

- *Subjectivism*—do what feels right, an inner focus notion. This emphasizes the person and so does not directly account for the feelings of others except as they affect the person's own sentiments. A similar idea is *intuitionism*—follow your instincts.
- *Divinitism (supernaturalism)*—do God's will, which perforce assumes it is discernible. Since different religions have different divinity structures and strictures, supernaturally focused societies often find themselves in conflict with each other and with those that don't sanction or are antagonistic towards particular religions.
- *Universal prescriptivism*—do what's approved in a fixed code of imperatives. This presumes that certain situations always call for the same ethical behaviors, leaving little opportunity to invoke extenuating circumstances.
- *Utilitarianism*—do what results in the best ratio of positive to negative outcomes, the greatest good for the greatest number. It assumes that the calculation can be made, which in turn assumes that possible outcomes are known in advance.
- *Consequentialism*—do what produces good outcomes, what has worthy consequences. This is a *teleological* approach, which is a more general idea than utilitarianism, but has the same focus on results.
- *Deontology* takes the opposite view of teleology. It eschews value judgments, holding that certain things are right and others are wrong regardless of outcomes. In some interpretations, deontology allows for choices being good if the person acted with good intent.

Conflict of interest   We depend on people in particular positions to do their jobs impartially. A conflict of interest can jeopardize that stance. Ethical codes usually call for people to avoid participating in matters where conflicts lie, or at least to acknowledge the conflicts. For example, a TV news reporter describing an event that concerns the parent company of the newscast network should disclose that fact; a judge should not preside in a trial in which he/she knows or is related to any of the parties involved; a journalist should not present reportage dealing with circumstances involving friends, relatives, or close associates.

The idea is that people in positions of trust should not take part in situations where their judgments could be unfairly influenced. But human nature is such that it is practically impossible to remain completely neutral in most situations. Whether consciously or not, people tend to interpret information selectively. Of course, we don't live in bubbles and can't help but have our viewpoints play a role. But there are some boundaries that shouldn't be crossed. Two examples:

- Prior to assuming her position as Supreme Court justice in August 2010, *Elena Kagan*[3] was appointed Solicitor General by President Obama in January 2009. In that position, she represented the federal government in cases that came before the Supreme Court and performed other legal duties on behalf of the government. In her Senate confirmation hearings, she responded to questions from Senator Patrick Leahy about conflicts of interest.

I would recuse myself from any case in which I've been counsel of record at any stage of the proceedings, in which I've signed any kind of—of brief.

In addition to that . . . I would recuse myself in any case in which I'd played any kind of substantial role in the process . . . I think that that would include any case in which I've officially formally approved something.[4]

Early in her tenure as a justice of the Supreme Court, over twenty cases came up that she had been involved with in one way or another as Solicitor General. Accordingly, she recused herself from those cases.

- *Michele Norris*, a journalist of some repute for over twenty years, hosted National Public Radio's *All Things Considered* news program beginning in 2002. Late in 2011, her husband, Washington lawyer Broderick Johnson, joined President Obama's 2012 campaign team as a senior advisor. Since Norris's job at NPR often involved reporting on presidential and campaign developments, she took a temporary leave from hosting the program.

Given the nature of Broderick's position with the campaign and the impact that it will most certainly have on our family life, I will temporarily step away from my hosting duties until after the 2012 elections.[5]

In these and similar situations, the choice to remove oneself from participating often includes instances where there may only be an appearance of a conflict though there is none. That becomes a fuzzy ethical line. People in sensitive or influential positions often choose to act as though the conflict was concrete. That may be an unfortunate consequence, especially if they feel that the taint of suspicion is as powerful as an actual conflict, a side effect of ethical pressure that can itself be unethical.

Then there are situations where a person's activities, whether questionable, unethical, illegal, or immoral, are separate from their official duties. That is sometimes used as a justification for staying in a job. But if those activities cast doubt on the person's integrity, character, and trustworthiness, to excuse them casts doubt on a person's code of ethics.

- In 2008, *Charles Rangel*,[6] a New York congressman since 1971, was accused of ethics and tax law violations. There were three main issues: using his office to raise money for the Rangel Center at the City College of New York, one of the senior colleges in the City University of New York; personally renting several rent-controlled apartments in New York City in violation of rent laws; and failing to acknowledge his income from a villa he owns in the Dominican Republic.

  Rangel fought the charges, but in March 2010 during investigations by the House Ethics Committee he resigned from his post as chair of the House Ways and Means Committee. Nine months later, the Committee found him guilty of violating eleven Congressional ethics rules. Finally, in December the full House voted to censure Rangel.
- *James McGreevey*,[7] who fathered a daughter in each of two marriages, began serving as Governor of New Jersey in January 2002. He resigned in November

2004 after he admitted to having an affair with a man and being gay. Was the issue his being gay, or was it being dishonest and unfaithful? Here is what he said: "Shamefully, I engaged in an adult consensual affair with another man, which violates my bonds of matrimony . . . It was wrong. It was foolish. It was inexcusable." He reasoned that he had to step down as governor to shield the office from "rumors, false allegations and threats of disclosure . . . removing these threats by telling you directly about my sexuality . . . I am required . . . to do what is right to correct the consequences of my actions."[8]

What do you think of this? *To recuse or not to recuse, that is the question*

On November 14, 2011, the US Supreme Court indicated it would hear a case brought by 26 states challenging the constitutionality of President Barack Obama's healthcare reform bill. Supreme Court Justice Clarence Thomas has been urged to recuse himself, since his wife, Ginni Thomas, founder of the conservative Liberty Central, is a constant fighter against what she calls Obama's left-wing agenda. Liberty Central has frequently advocated against the reform bill.

Although Thomas's wife no longer heads the organization, she has continued to lobby against the healthcare bill. So while Thomas has no direct fiduciary interest in Liberty Central or other healthcare lobbying groups, which would be an immediate cause for recusal, is Thomas ethically bound to recuse himself because of his wife's activities? Does the fact that between 2003 and 2007 Ginni Thomas added over $680,000 to the Thomas household income, payments she received from the Heritage Foundation, a conservative group that opposes the healthcare reform bill, make a difference?

Most of the recusal urgings come from those supportive of the healthcare bill—primarily liberal organizations and Democrats, including a letter signed by 74 House Democrats. Most of the opponents of recusal come from those opposed to the bill—mainly conservative organizations and Republicans. They also claim that Justice Elena Kagan should recuse herself because she used to work for the Obama administration. Does Thomas's stance as a conservative justice mean that he could not be impartial in this case? Does Kagan's past role in Obama's administration mean she could not be impartial? What ethical considerations should apply?

It's apparent that ethics has many dimensions, not only in philosophical viewpoints but also in people's perspectives and differences. Keeping this book's emphasis on IT in mind, we leave our general overview of ethics to the foregoing and focus the remainder of the chapter on the intersection of ethics and IT.

IT enters the picture

As computer-based IT sprang to life, we found ourselves with increasingly more ways to communicate, to collaborate, to utilize new sources of information, and to have more influence than was previously possible. Now there's no getting away from digital IT. It's part of our everyday lives, whether we use it directly or it operates in the background.

We have in our hands possibilities for actions and interactions that didn't exist before. We've moved from a time when we mostly acted locally to global communication as the norm. And as IT has shaped the world we live in, our responses have shaped IT.

That poses a question: how do we choose to operate in this milieu? We can say let ethics be our guide, but which ethics, whose ethics, and do they apply to today's world? Ethical behavior in one culture is not necessarily ethical in another, so moving to a global perspective can leave us on ethically shifting ground.

Another gap in the bases for ethical decision making comes from the disparity in the rapidity of technological change and the slowness of legislative action. That can leave us unsure of what to do, which can lead to using IT in ways that can be illegal but justifiable on technical grounds. If it can be done, why not do it? That's a subjectivist's ethical choice.

IT as a neutral actor   A common notion is that technology in general and IT in particular are tools that have no inherent ethical dimensions. Most technology can be used in more than one way, in concert with or divergent from original intent. From that perspective, technology is neutral—what we do with it is up to us. We can use a hammer to build something or to knock someone over the head. IT is no different. We can send email to a friend or we can send spam to millions.

Even technology that by design is not neutral at all can be viewed from both perspectives. Bombs are an example. They are designed to destroy, but we can channel their use positively as deterrents by threat of their use. And in accordance with consequential ethics, we can justify bombing an enemy to bring an end to a conflict sooner than otherwise, thereby saving more lives than are ended by the bombing.

Again, IT is no different. Malware is designed to do nasty things. What can be said about it that's positive? Malware can deter cyber warfare by threat of retaliation and it can be used preemptively to stop a bad actor from proceeding, a consequentially ethical decision.

In all these cases, extreme or not, we bear a measure of responsibility. How does our society view injustices? Do we punish, do we educate, do we retaliate, do we negotiate? Valid ethical points can be made in support of any of those actions. It is up to us to inform the decision makers of our ethical views. In democratic societies that's done by corresponding with our representatives, by forming action groups to spread our sentiments, by voting for those who share our opinions, and so on. In all those endeavors, we can enlist the power of IT to amplify our messages.

IT as a preordained actor   Any given IT, like any tool, has a designed purpose that eventually will rise to the fore. Once the capability exits, it will be used in all sorts of ways, ethical or not. For example, data mining is designed to seek patterns in massive amounts of data that would otherwise go unnoticed. Businesses use data mining effectively to better understand their markets and operations. They also use it for targeted advertising that, in the process, can invade our privacy. Spyware is designed to gather information covertly. It can be justifiably used to track criminal activity, but it has found its way into our personal computers to snag our account numbers and passwords.

In these and similar instances, the key is the usage, not the tool. To abandon or prohibit tools that can be used malevolently means losing the good and proper usages they provide. The ethical choices to be made involve understanding how a tool's design can contribute positively and negatively and then deciding to use it or not. We know, for example, that we can trade privacy for convenience and that is our choice. But how are the organizations that gain our private information acting? Do we consider their behavior ethical? The answer should play a part in our decision to participate. In the end, we may have to choose between convenience and ethics.

IT as an ethical actor   IT, ethics, and morality are inextricably bound, so it follows that IT products and processes cannot be value free. Sooner or later, especially as their influence spreads, ethical issues come to light that demand attention. We've found that unforeseen gaps in legal and behavioral codes can leave us without clear guidelines in resolving those issues on both individual and societal levels.

Value sensitive design takes it as given that no technological design is value neutral, but perforce carries ethical and moral implications. Rather than ignoring them, they should be a conscious part of the design process considered very early on and not left to happenstance.

Following this precept means resolving conflicting ideas about which ethical and moral principles are to be considered in a given situation. While an optimal resolution may not be possible or even definable, taking the issues into account has to result in designs that at least have a chance of being socially positive. This concept can be applied in many arenas by establishing socially relevant and responsible decision making as a goal, whether applied to devices, computer programs, legislation, or policies. So we can readily relate it to IT development.

Value sensitive design fits with the notion, noted in the previous section, that any IT has a designed purpose that eventually will rise to the fore. If that purpose incorporates ethical and moral principles, then we can expect that the chances of it being used in a socially responsible way are increased. Of course, there is no certainty that those principles will exert a positive influence, nor is there any reason to believe that users can't employ it in unintended, negative ways, but at least the odds swing towards the positive.

What do you think of this? *Bugs on board*

It's not unusual for software products to come with bugs, most of which are eventually addressed with patches and new releases that themselves can contain bugs. All the while, consumers serve as testers, a shadow cost in the form of free labor for product developers. So why are buggy products released?

Many programs have become so large, not to say bloated, that it's all but impossible to check every aspect of their operation to sniff out bugs. Even for simpler software, bug free the first time around is a rarity. It happens because in the highly competitive world of software, companies want to get their offerings to market without delay. That leaves only a limited amount of time for checking program code.

So, is there a level of bugginess that's acceptable? How quickly must producers respond to errors? And the grand question, is it ethical to market buggy software?

Funding influences development According to *technological determinism*, technology follows its own progression independent of societal influences. That is, societies and their cultures are pushed in directions consistent with technology's mandates. There's no doubt that technology is a factor in shaping societies and their norms, but is it just a one-way street?

Most IT doesn't come cheap. Despite rapid advances, development cycles can be long and lifespan can be short. Funding is a critical element in supporting that kind of operation. Potential developers have to attract funders to back their projects. Funders independently dangle the prospect of support for specific projects related to their own goals. Either way, funders have considerable influence over what gets researched and developed.

The influence of governments is tremendous, from specific inner-directed concerns to such broad and far-reaching areas as social welfare, world status, and influence. Shifting political sands mean shifting intent as politicians and government leaders attempt to discern and to shape public opinion. So we find money variously flowing into areas as diverse as broadband development and access, competitive communications services, analog to digital conversion projects, wireless spectrum efficiencies and allocations, satellite surveillance networks, information security and processing, and cyber warfare defense and offense.

A classic example of government's influence on IT development is the project funded by the US Department of Defense (DoD) that created the ARPANET, which evolved into the Internet.[9] Naturally, the DoD had specific defense-related objectives in mind that directly influenced the work.

Self-funded projects by deep-pocket companies are no different. Their goals are influenced by competition, the marketplace, feedback from customers, shareholder equity, and stakeholder valuation, so they prioritize outlays and development directions accordingly. Private equity sources, venture capitalists, angel investors—all look for projects that fit their stated missions while minimizing risk and maximizing return.

How much do such ethical issues as reducing harmful emissions, decreasing waste byproducts, and aiding particular groups influence investment decisions? After all, even socially conscious investors and developers must aim to create sustainable businesses supported by appropriate revenue streams. How much weight to give social benefit in business investment decisions is a question of balance, something that's becoming a factor in all sorts of business decisions. When practicality enters the picture, that can be a difficult line to draw.

Intersections

In previous chapters we explored some of the issues that arise when we're faced with conflicting objectives. We've also seen the impact that legislative lag can have on IT implementations, but it's not rational to expect IT development to slow down so the law has a chance to catch up. If we require it to, then development would have to proceed in fits and starts—hardly a viable solution.

We noted that value sensitive design encourages injecting ethical considerations at the start of the design process, but it's a rare product or service whose ethical dimensions are fully apparent before the IT is put to use. That's because it's not realistic to expect that any IT can exclude all but originally intended usage and it's impossible to divine everything a developing technology will eventually be turned towards.

Ethical considerations can clarify or muddy the picture. Still, when it comes to creating and using IT, taking ethics as a guide at least gives us a leg up on acting appropriately. That, of course, is easier said than done.

Free expression   Written into the First Amendment to the US Constitution, freedom of speech has morphed into freedom of expression. Self-expression is taken to be a fundamental right, whether in the form of actual speech, writing, music, art, or almost any other way to demonstrate ideas and opinions.

Speak out against the government and politicians, including the president? Participate in the Occupy Wall Street movement and demonstrate against corporate officers, bankers, financiers, and financial firms? Stand on a street corner and wave signs urging us to repent as the end of the world is near? Burn the flag? Submerge a small plastic crucifix in the artist's urine? Protest the government's funding of controversial art? Support family planning? Object to abortion? Yes to all and more, yet there are limits.

One has to do with venue—owners can restrict what happens on their private property. Another has to do with consequences—does a demonstration pose a danger to participants and bystanders or damage property? Then there are legal issues—libel, slander, and a plethora of laws proscribing particular behaviors that can't be violated with impunity.

For the most part, exceptions are clear, but what happens when exercising the right to express oneself conflicts with someone else's rights? Does the noise of a demonstration take precedence over the right to peace and quiet of people living close by? Is spam free expression or invasion of personal space? Does denying marchers a permit inhibit their right to self-expression?

Leakers   Disclosing private information is unethical, or is it? When whistleblowers expose fraud, malfeasance, product defects, falsified test results, or similar corporate or government misdeeds, they can be considered to be providing a public benefit. Therefore their behavior is not only ethical but morally required. Yet the virtue of such disclosures is not always easy to judge.

When WikiLeaks posted leaked US diplomatic cables, was it an act of treachery, perhaps even treason—unethical and illegal—or was it a revelation of improper behavior and cover-up by diplomats—ethical and protected? The problem is, once information is leaked, it quickly becomes globally available whether or not the action was valid or commendable, whatever the intent, and regardless of consequences.

Pornography   One of the most controversial areas of expression, pornography as an adult activity is generally a protected form of expression, but the Internet has complicated the picture. Localities can pass laws restricting where so-called adult entertainment businesses can operate and what defines an adult. Patrons of establishments can be checked to see if under-age people are being admitted.

Restricting Internet distribution of pornographic material is another matter. Age-related limitations have been legislated to prevent access by juveniles. To comply, many online sites require a viewer to indicate that he/she is at least of legal age, but there are no foolproof ways to prevent anyone from lying. Many free sites don't require any form of ID and having to provide one could be considered an invasion of privacy.

Child pornography is another story altogether. Most countries have an abiding ethos against any of its forms. Laws prohibiting not only distribution, but also possession of child pornography are widespread and nearly every country cooperates in pursuing and prosecuting violators. Yet it is a multibillion dollar worldwide industry, largely enabled by the Internet and the difficulty of tracing sources, that shows no signs of abating, ethos or not.

Children and pornography is somewhat different. There the emphasis is on protecting children from exposure to pornography of any sort. In the US, the *1996 Communications Decency Act* (CDA), incorporated in the *Telecommunications Act* of the same year, was intended to do just that and more. It was a broad stroke attempt to criminalize portraying so-called indecent material on Web sites, so broad that a lawsuit challenging its constitutionality was quickly brought.

Led by the American Civil Liberties Union (ACLU), the Electronic Privacy Information Center (EPIC), and 18 other disparate organizations, *ACLU v Reno* declared CDA unconstitutional. Brought on appeal by the Department of Justice to the Supreme Court in what was renamed *Reno v ACLU*, the ruling was upheld in a 1997 decision declaring free expression to be a protected right afforded on the Internet just as it is for any other means of expression.[10]

Harmful material   Still focused on protecting children but reaching beyond pornography, Congress passed the *Child Online Protection Act* (COPA). Signed into law in 1998, it required all operators of commercial Web sites to restrict access to what the law called material harmful to minors, the definition of which was supposed to be based on community standards. By restricting the scope of the law to commercial activities and US-based sites, the hope was that objections to the broadness of the CDA that led to its demise would be overcome.

However, the law was so vague as to be interpretable at will to suit just about anyone's idea of what was harmful or not. The law was quickly enjoined, a block that was upheld on appeal to the Third Circuit Court in 1999. Appealed again to the Supreme Court, it was returned to the Circuit Court in 2002 for further review. Once again, in 2003 the ruling was upheld.

Far from the end, additional appeals traveled up and down the court systems, where it met repeated confirmations that the law was unconstitutional. On its last trip to the Supreme Court in 2009, the case was refused, and so COPA was finally laid to rest.

Not wanting to wait for the outcome of the COPA appeals' journeys, Congress made another attempt at legislation. Instead of trying to rein in pornographic sites or other harmful content providers, the idea was to block access to those sites. The *Children's Internet Protection Act* (CIPA), passed in 2000, required libraries and schools that received federal funding to install filters on public access computers that would prevent viewing such sites.

CIPA didn't specify how to implement site blocking or what content to block, leaving that to the individual libraries and schools. Some took the low-tech solution of placing computer monitors where they could easily be seen by adult supervisors. Others tried various software filters. Still others objected to the law as an unnecessary restriction of First Amendment rights that also could prevent well-intentioned and important information seeking about such topics as drug abuse, sexually transmitted diseases (STDs), and the like.

CIPA followed a different course in the courts. In 2002, an Eastern Pennsylvania panel of three judges concluded that it forced libraries to violate the First Amendment rights of their patrons and so was unconstitutional. However, an appeal to the Supreme Court led to overturning that verdict. That conclusion was "public libraries' use of Internet filtering software does not violate their patrons' First Amendment rights, [so] CIPA does not induce libraries to violate the Constitution, and is a valid exercise of Congress' spending power."[11]

The Court further concluded that protecting minors from harmful material was appropriate and as long as librarians could provide unblocked or unfiltered material to adults on request without delay there was no First Amendment issue. Librarians still were left with considerable leeway. For example, they could designate certain machines as adult only or, as noted above, put monitors in plain view of librarians who could see if minors were using the computers inappropriately. Of course, the issue of what constitutes harmful material was not settled, nor was it mentioned in the rulings except for pornography.

Deception   Email is an obvious medium for speech, but should its use be unfettered? When email is uninvited, is it ethical to stop it, or is it only an issue when it deceives? In 2003, the *Controlling the Assault of Non-Solicited Pornography and Marketing* (CAN-SPAM) Act addressed those subjects by bringing purpose into the equation.

According to the Act, email addresses and domain names have to accurately identify the sender; the subject line has to be clear as to intent; ads have to be plainly noted as such and contain actual return addresses; and recipients have to be able to opt out of future mailings. Unfortunately, the Federal Trade Commission, charged with enforcing the Act, has not been very active in doing so although there have been some fines levied and lawsuits brought.

Defamation of character is a free expression exception addressed by libel and slander laws as actionable offenses. The courts are the ultimate venue for settling disagreements as to whether a disputed expression was legal. But what about more subtle deception that could cross over to abuse of free speech? Software for manipulating digital images and videos lets us be remarkably creative. Is such falsification artistry, ethical and protected speech? Suppose the aim is to obfuscate, mislead, or outright deceive, such as doctored before-and-after pictures used to market a wrinkle cream—still ethical?

These software products are examples of IT tools whose use cannot be the responsibility of their producers. But to act ethically in the name of transparency, should they incorporate means for detecting when an image has been modified?

Gaming the systems   Ethically questionable though perfectly legal behavior has come about as a reaction to the influence of search sites like Google, Yahoo, and Bing. Where a business's link falls on a list of search results can have a significant impact on its bottom line, leading some to seek higher placement by taking advantage of the way search site ranking algorithms work. Some businesses offer services to improve rankings; some businesses hire people to repeatedly link to their sites to raise popularity, hence list placement.

The search engine battle is in full force. Opposing manipulation attempts, search service providers modify their algorithms to try to keep pace so as to maintain control. Is that for our benefit or for their own? Is either side acting ethically?

The recommendation game is similarly in play. Most sites that sell merchandise and services provide user-submitted comments, good, bad, and indifferent, along with their offerings, but don't vet the submitters. There is no way to know if the commenters are at all reliable, or even what motivates someone to leave a comment or review. Are there any conflicts of interest?

Privacy   Is there anything we do that isn't recorded somewhere? Doubtful. Are secrets a dying concept? Quite possibly. The weight given to privacy is declining in tradeoffs between it and convenience. Moreover, loss of privacy is increasingly viewed as an inevitable consequence of the digital/Internet age. Accordingly, the ethical significance of privacy is losing ground.

Our online and phone activity, our movements, whom we associate with, and more are goldmines of information for whomever can take advantage of them, with and without our knowledge and with and without our consent. Is unconsented collection ethical? What about unconsented use?

We could say that improved search results, more accurately targeted advertising, easier associations, and the like are ethical because in sum they are beneficial for more people than not. Of course, that's hard to prove, but assuming it's true there still is a balancing question—is it simply a matter of the greatest good—one ethical model—or are some negatives, even if only for a few, enough to tip the scales? A question to ask is, what is the potential for invasion of privacy?

Identifiability   Privacy issues revolve around personally identifiable information (PII)—who collects it, who has access to it, who has control over it. Some such information we have to part with and always have had to. The Internal Revenue

Service knows a lot about our finances, as do banks and other financial institutions. Our addresses and phone numbers are no secret. Birth records, home ownership, property tax, home sales, rentals, and similar information are a matter of public record.

We give up some PII willingly and knowingly—when we make purchases online or in stores using credit cards, debit cards, and customer loyalty cards; when we apply for a passport and travel abroad; when we take a flight to anywhere. Still other information we purposely supply—when we post on social network sites, rate and describe experiences with products and services; email a letter to the editor of a newspaper or comment on a blog post.

On the other hand, there is a big difference between what is collected and how it is used. Companies have taken advantage of online sources for background checks of potential and current employees. As we saw in Chapter 4, you may be turned down for a job or fired from one because of something you posted on a social network site that has no reasonable relevance to work. Of course, you are responsible for what you post and should be quite circumspect in your choices. Though viewers, whether employers, friends, acquaintances, or anyone else, have an ethical obligation to treat your posts appropriately, you can't rely on their discretion. Do you have an ethical obligation not to put temptation in their path?

What do you think of this? *We know who you are*

In the Chapter 4 section called Identifying the unidentified, we noted: "Information that has been anonymized by removing such personally identifying data as names, social security numbers, and addresses can often be de-anonymized to reveal the person behind the data." Is it ethical to publicize techniques for de-anonymizing data? Is it ethical to de-anonymize data? Are there circumstances when it is?

Surveillance   Another form of data gathering, surveillance can be legally and ethically undertaken by police and other enforcement agencies. However, the capabilities of IT have raised issues that have yet to be resolved. Consider GPS. It's a simple matter for a device to be planted on a car without knowledge of the owner and then used to track his/her travels. Details of trips over several months can reveal comprehensive patterns of routes taken and places visited that can be incriminating, perhaps justifiably so, perhaps not. When undertaken without first obtaining a warrant, is that an invasion of privacy, a violation of Fourth Amendment unreasonable search and seizure?

The usual justification for GPS tracking is that it's simply a technological extension of physically tailing someone, which has long been a routine part of warrantless police work. In comparison, the likelihood of a physical tail producing as much data and being successful over an extended period of time is much smaller and the cost surely much higher than following an electronic trail. So is claimed suspicion enough for an agency to move on GPS tracking or is court oversight ethically, if not legally, appropriate?

Cameras mounted in public places, in stores, in banks, in buses and trains, in police vehicles, in corporate facilities, put us in view almost everywhere. They have proved useful in identifying and finding criminals, in sorting out conflicting descriptions of events, and as a deterrent to crime. At the same time, that amazing loss of privacy has become so routine as to have receded from consciousness. Has that lapped over to other areas of privacy loss, pushing privacy protection farther below other considerations?

What do you think of this? *Good use of dubious IT?*

In Chapter 7 we saw how spyware and key loggers can steal our data surreptitiously. It seems safe to say that spreading and taking advantage of such malware is illegal and unethical, but is it always? What if it's used by the FBI, the CIA, the NSA, or similar agencies to catch criminals or stop terrorists? Would obtaining warrants first make a difference or are such methods always unethical search?

Transparency   Although we know that privacy is not what it was and many of us feel that its continued decline is inevitable, we may not realize the extent to which it has waned. Importantly, we rarely know what's done with our data and how that affects our lives in myriad subtle ways. What if we did know?

Suppose data collectors, data manipulators, and data distributors were open about their practices? That would certainly raise our awareness. As a consequence, it might move us to think more about what's going on and lead us to reevaluate how we value privacy.

Implications   Data collection and storage of all sorts of information in vast databases is continuous, thorough, and far reaching. Multiple instances of the same data are common, whether or not those entries are erroneous. Without rigorous data cleansing, conclusions drawn and actions taken can be wrong even to the extent of being harmful. Think, for example, of how errors in financial data can damage a person's credit rating. After all, data mining does not delve into why a pattern pops up, only that it does.

Of course, mined information can be put to reasonable and proper use—to improve our shopping and search experiences, to simplify form filling and filing by remembering what we've entered before, and by making it easier to associate with people with similar interests. That makes data mining seem benign, but there are legal and ethical questions to consider.

If information is taken without our knowledge or consent, does that violate the basic tenet of the Fourth Amendment's proscription of unreasonable search and seizure? Does providing us with some way to agree to (opt in) or prevent data capture (opt out), however arcane, relieve sites of responsibility? Should we be able to see what's been collected about us, to correct or delete information, to easily restrict its use? Can we expect data to expire after a limited specified time? Do claims of data anonymization give companies a pass? And the overarching question: to whom does our data belong? The answers are a mixed bag.

The Internet and digital IT have produced seismic changes. Digital data easily riding on the Internet can neither be contained nor retrieved. That makes global solutions impossible and national solutions near impossible. While we may seem to be giving up only local privacy constricted to a specific site, we don't control how that data travels.

Yes, some sites give us more control than others and some make it easier to exercise that control than others. And yes, changes have come about as we have become more aware of what's going on and legislators have responded to pressures to do something about it. So particular kinds of sites have been forced into particular kinds of compliance—mostly minimal, typically local, but at least a start. Still, in the final analysis, how the data mountains are manipulated and used is largely up to those who have access to them.

Intellectual property   The ease with which digital information can be manipulated has rendered some forms of protection for intellectual property (IP) moot. Ethically suspect arguments have surfaced that represent a shift away from the idea of ethical behavior and towards the pragmatic—if the tool is there, use it. Capability trumps ethics and the law.

A prime example is the huge amount of illegal downloading of music and videos, not possible before digital releases and broadband Internet connections. Perpetrators rationalize their actions by saying that music and videos are too expensive to purchase, free downloading is universally available, it's easy to do and everyone does it, so why not?

Due to legislative lag, anti-piracy laws took a long time to develop. By then, illegal downloading was so ingrained as to become a cultural norm for particular societal subgroups and so, by one definition, ethically consistent behavior.

The concept of theft vanished from the equation. Yet the great majority of the same people are not inclined to steal a car whose owner left the keys in the ignition, swipe a handbag that's hanging over the back of a chair within easy reach, or lift a few bills out of the register of a distracted cashier. If the same "why not" rational was applied, there would be hugely more auto thefts, stolen handbags, and disappearing cash. Is the difference online rather than in-hand, digital rather than physical? The morality applied to the Internet seems to be different.

Copyright and patent   The idea of copyright is quite straightforward—give the holders of expressed ideas property rights and creativity will be encouraged by the rewards that ownership can bring. The key word is *expressed*—as we saw in Chapter 5, it's the expression of the idea that's copyrightable, not the idea itself. Applied to literary works, art, music, and the like, it's usually easy to understand what expression means. The situation with software is another matter.

Currently, computer programs are copyrightable, but the underlying algorithms and procedures that are the heart of the programs, even though expressed in the software, are not. That peculiarity means that reverse engineering program code to learn how its processes work and then expressing them slightly differently in new programming is not a violation of copyright since the new program is different.

A separate question is the length of copyright protection, now 70 years past the life of the author and 95 years after first publication of anonymous work or work for hire (120 years from first expression if that's sooner). That stricture goes far beyond what's required to encourage creativity.

Patents follow the logic of copyrights—give inventors property rights and innovation will be encouraged. The term, just 20 years, is more reasonable since most inventions usually have outlived their usefulness by then anyway, having been superseded by improved versions. That prods patent holders to capitalize on their work quickly and to keep control by continually making design changes sufficient to garner new patents. So innovation is encouraged, as is the value of a patent.

Patent value has stimulated a different sort of enterprise—businesses that buy patents from bankrupt and cash-hungry companies solely to accumulate portfolios that can be used to threaten others with patent violation lawsuits. The mere threat of a suit often is enough for the victim to capitulate by paying fees or ceasing use, because the expense of fighting a lawsuit can be overwhelming. These so-called patent trolls operate legally, at least for now, but hardly ethically.

Patents do provide another possibility for programmers. While copyrights protect expression, patents protect function. That might seem to alleviate the algorithm problem, but mathematical algorithms can't be patented, so can any algorithm be? As it happens, several thousand software patents have been granted despite a 1981 Supreme Court decision denying one, although the 5 to 4 vote was hardly a ringing testimonial.

Legal and technical issues can be argued back and forth, but they beg the ethical question—is it right to copy someone else's work and use it to your own benefit? That's hard to justify.

What do you think of this? *Ethically justified activism or scare tactics?*

On Saturday, January 21, 2012, faced with the prospect of an impending vote on the House of Representative's *Stop Online Piracy Act* (SOPA), which many believe is an unnecessarily heavy handed approach to reducing IP theft that was strongly influenced by movie and music producer lobbies, WikiPedia took the drastic step of blacking out their site for 24 hours. A query to the site led to a black and gray page with an ominous statement that included the following wording:

Imagine a World Without Free Knowledge

Right now, the US Congress is considering legislation that could fatally damage the free and open internet.

Other sites took less drastic measures, adding messages to their pages asking people to complain to their representatives. Still, some of their statements were over the top, calling the SOPA and its Senate parallel *Protect Intellectual Property Act* (PIPA), attempts to censor the Web.

Hyperbole in the service of getting a point across is a common tactic. In this case it worked well, as millions of people signed petitions and contacted legislators in protest. Under the deluge, both bills were put on hold. But the actions by the sites raises a question: Is exaggeration to an unrealistic level, aimed at frightening people, ethically suspect?

The bills did have problems, including some ambiguous language and questionable provisions that could have resulted in inappropriate restrictions on Web sites and ISPs. On the other hand, though not the whole solution, can IP theft be reduced without the aid of legislation? Did the protests put legislators in the position of fearing to proceed with any attempts to address the problems of IP theft?

Artificial intelligence

The greater the freedom of a machine, the more it will need moral standards.[12]

In Chapter 8 we saw that artificial intelligence is not a monolith. It comes in many forms, has many goals, and many beneficial uses. Where discussions about possibilities often fall apart is when it comes to how far machine intelligence can go and how far it should go.

Seeking true human-like AI is denigrated by some of its detractors as a fool's errand because they do not believe any machine can have such presumably unique human characteristics as a conscience, a sense of responsibility, or an intuitive grasp of right and wrong. Consequently, they fear that intelligent machines, unerringly moving towards an action, a performance, a mission, might unwittingly cause enormous harm.

People, on the other hand, by progressing tentatively when facing fuzzy boundaries, conflicting outcomes, and ethical uncertainty can come to better conclusions than unswerving machines. Yet ironically, there's something about the thought of a dithering, more human-like machine that's even more unnerving than an unwavering one. Then too, the history of dismal human behavior can put the lie to the idea that it's a good idea for machines to conduct themselves more like humans.

What do you think of this? *Human labor, machine labor*

The industrial revolution ushered in machines that grew increasingly capable of performing manual tasks that were done by people. A great hue and cry arose as workers were displaced by machinery. Although the short-term disruptions in the labor market were painful for the displaced workers, society as a whole benefited from increased productivity and the birth of new businesses that built and serviced the machinery. The model of temporary dislocations in support of the greater good came to be viewed as the right thing, the ethical thing.

Workplace displacement is happening now too, but with a big difference. Instead of so-called blue collar workers bearing the brunt of labor dislocations, it's moving to mid-level white collar workers whose intense repetitive work is being taken over by AI-enabled machines. Does that change the picture? What if replacement moves to upper-level white collar workers?

Suppose we do reach a point where machine brains have all the capacity of human brains. Will ethical awareness and a conscience naturally follow? Could such a machine learn ethical behavior as humans do? Would it also waver in the face of uncertainty, perhaps making the wrong choice as we sometimes do? And if so, would it be responsible for its actions, tried by a jury of its AI peers? The grand question: Is it possible to build ethical, moral machines?

Seventy years ago, Isaac Asimov took a crack at writing a code for AI machine (robot) behavior. His three laws:[13]

1. A robot may not injure a human being or, through inaction, allow a human being to come to harm.
2. A robot must obey orders given it by human beings except where such orders would conflict with the First Law.
3. A robot must protect its own existence as long as such protection does not conflict with the First or Second Law.

These laws were the basis for many science fiction novels that dealt with the interaction between humans and robots and the struggles to prevent robots from taking over. As an ethical code, though, it comes up short. For example, the First Law would stop a robot from injuring a felon it was pursuing. If the felon and the police were shooting at each other, whose harm would take precedence? And if the felon ordered the robot to help him escape, should it be obeyed as per the Second Law?

Asimov was not trying to create a code of robot behavior that would parallel a human ethical code. Taken for what it is, it focuses on keeping robots a notch below humans on the behavioral scale, holding them subservient. Going a step farther, in 1985, Asimov added what he called the zeroth law, which superseded the others. Interestingly, it took the view that societal good supersedes individual importance, placing concern for humanity above that of an individual:[14]

0. A robot may not injure humanity, or through inaction, allow humanity to come to harm.

Two other laws, called the fourth and fifth, were proposed by others to help in keeping the proper balance between robots and humans:

4. A robot must establish its identity as a robot in all cases.[15]
5. A robot must know it is a robot.[16]

Other laws have been proposed, all with the gist of making sure humans reign supreme.

To make the laws functional they must be embedded in robot brains, in robot psyches, so to speak. Then ethical behavior will follow inevitably. That could be a starting point for creating ethically acting AI machines. There is another issue, however. Codes of this type leave out the behavior of humans who use or interact with robots. Since unethical behavior is all too common among people, there's no reason to assume that people would act ethically towards or with robots.

The problem is, humans or machines strictly following any sort of behavioral code will inevitably come across ethical dilemmas for which a code doesn't suffice. As humans, we may rely on such Aristotelian concepts as virtue and character to guide us, and as humans, we can fail.

It's not easy to develop virtue and character, but most of us do, though none of us is perfect in that regard. Imagine how hard it is to build machines to do likewise. But hard doesn't have to mean impossible. So suppose we can build fully AI machines that develop virtue and character and guide their actions accordingly, supported by, but not chained to, ethical codes. Just like humans, at times they will fail too. Are we prepared for that?

An easy way out is to answer negatively the question: Do we want AI machines to be able to make decisions that have ethical consequences? As IT and allied technologies march on, that development might very well be inevitable anyway. Philosophers have engaged in the topic of ethical behavior for centuries. Since it's always easier to address issues at the design stage than afterwards, should AI design teams include ethicists, both to instruct designers in value sensitive ways and to ensure that appropriate innate machine behaviors are part of robot intellect?

That might work well if there was agreement as to what ethical behavior is, human or machine, especially in fuzzy situations. Unfortunately, there isn't. Moreover, it's not possible to anticipate all effects at the design stage. Unintended and unanticipated consequences always emerge. That's why software comes in versions, why patches are issued, and why product recalls to correct flaws are instituted.

In any event, we're back to the question of certitude—do we want machines to dither in ambiguous situations as do people? If not, they need to have definitive behavioral algorithms that will cause them to blunder ahead to bad results at times—no code can cover every situation perfectly. Yet if we do want them to muddle through foggy terrain, they may choose a path that takes them off a figurative cliff. Either way, intelligent machines are in an untenable position—when forging ahead no matter what or when choosing among uncertain outcomes turns out wrong, judgment goes against them—in a word, human-like.

What do you think of this? *Imitating humans*

If machines can think as humans do, will they get bored if they don't have enough to do? And will that result in mischief or destructive behavior?

If they face dismal situations, will they get depressed? And if they do, will they metaphorically pull the covers over their heads and withdraw?

Will machines react to praise, modify their behavior to seek rewards? And if they do, will they become sullen if they feel overlooked or mistreated?

If machines are productive members of society, will they expect to get paid for their labors, get raises, participate in unions, have a say in management, become managers, create and own businesses? And if they do, will they act in concert with humans or compete against them?

Professional ethics

Most professional organizations have ethical codes that define appropriate behavior for members of their professions. They are similar in that while they espouse conduct specific to their fields, they are based on general ethical principles. A typical example is the *ACM Code of Ethics and Professional Responsibility.*[17] Among its several sections are these:

- *General Moral Imperatives*, describing how to: "Contribute to society and human well-being; Avoid harm to others; Be honest and trustworthy; Be fair and take action not to discriminate; Honor property rights including copyrights and patent; Give proper credit for intellectual property; [and] Respect the privacy of others; and Honor confidentiality."
- *More Specific Professional Responsibilities,* covering such areas as quality of work, professional competence, professional reviews, honoring contracts, and avoiding unauthorized access.
- *Organizational Leadership Imperatives*, discussing how to manage people appropriately, support their work, enhance quality of working life, and respect their needs and dignity.

The IEEE code of ethics[18] has similar provisos, succinctly stated in ten points:

1. to accept responsibility in making decisions consistent with the safety, health, and welfare of the public, and to disclose promptly factors that might endanger the public or the environment;
2. to avoid real or perceived conflicts of interest whenever possible, and to disclose them to affected parties when they do exist;

3. to be honest and realistic in stating claims or estimates based on available data;
4. to reject bribery in all its forms;
5. to improve the understanding of technology; its appropriate application, and potential consequences;
6. to maintain and improve our technical competence and to undertake technological tasks for others only if qualified by training or experience, or after full disclosure of pertinent limitations;
7. to seek, accept, and offer honest criticism of technical work, to acknowledge and correct errors, and to credit properly the contributions of others;
8. to treat fairly all persons regardless of such factors as race, religion, gender, disability, age, or national origin;
9. to avoid injuring others, their property, reputation, or employment by false or malicious action;
10. to assist colleagues and co-workers in their professional development and to support them in following this code of ethics.

Formal ethics codes and informal communal codes set the tone, but unless people follow through they are meaningless. While we need to buy into them and practice them, that doesn't mean we can't disagree on what's appropriate in particular situations or debate choices. It doesn't mean we won't make mistakes. But it should dissuade us from being the kind of bad actors who put themselves above the regard for the damage they inflict on others in pursuit of their own self-centered goals.

The green way

IT is a resource-intensive enterprise, beginning with the production of equipment, proceeding through usage and service provision, and on to disposal of electronic waste (e-waste). Most IT apparatus contains toxic elements that make proper handling, whether by recycling or final disposal, difficult and expensive. The amount that needs to be dealt with annually is measured in many millions of tons worldwide.

E-waste is a global problem and how we deal with it is perhaps the most blatant IT ethical issue that we confront. Much is made of the fact that developed nations are known to send significant amounts of their e-waste to China, India, and Ghana, among other countries. There, the marginalized poor, struggling with few if any earning options, process the material under primitive and hazardous conditions. That not only endangers the health of the workers, it also poisons the water, soil, and air, affecting everyone in the area and beyond.

Though significant strides have been made in dealing with toxic e-waste, a lot is still dumped on people who have little choice in handling it. But consider that those operations are not unknown in their countries. Then add that electronic equipment from those very countries is sent to the same facilities, and that exploited workers are not operating on their own—they are employed by those who take advantage of their dire straits and profit greatly from their labors.

So we must conclude that these operations are condoned by those countries' governments and local enforcement agencies, where political corruption and bribery hold sway. Though that does not excuse manufacturers and e-waste exporters, the ethics and responsibilities of the e-waste challenge redound as much to those who accept the waste and exploit their own people as it does to the producers and transporters of the material.

Most countries have signed onto the Basel Convention,[19] an agreement describing proper handling of e-waste, particularly addressing its export and import. Other countries have formulated their own similar regulations. Even the countries with the primitive recycling operations have laws contravening accepting exported e-waste, though it's obvious that those laws and regulations are routinely disregarded.

Several groups are working to rectify the situation. Prominent among them is the *Basel Action Network* (BAN) "the world's only organization focused on confronting the global environmental injustice and economic inefficiency of toxic trade (toxic wastes, products and technologies) and its devastating impacts."[20]

Another is *Greenpeace*,[21] a non-profit organization that addresses many environmental issues. One of them is promoting green technology, which looks at equipment from manufacture to disposal. They publish a *Guide to Greener Electronics*, updated every three months, that ranks "the 18 top manufacturers of personal computers, mobile phones, TVs and games consoles according to their policies on toxic chemicals, recycling and climate change." Rankings are based on

> the demands of the Toxic Tech campaign to the electronics companies. Our three demands are that companies should:
>
> clean up their products by eliminating hazardous substances;
> take back and recycle their products responsibly once they become obsolete;
> reduce the climate impacts of their operations and products.[22]

Green technology looks at the entire life cycle of IT products and services, from production to disposal. Attention is directed to production methods that are energy efficient and minimize emissions, and to end-of-life processes that recycle or properly dispose of byproducts.

What do you think of this? *Environmentally and monetarily green*

Services like search engines, information Web sites, cloud computing providers, and the biggest online merchandisers run enormous data warehouses and communications networks. Their computers and storage devices consume huge amounts of electricity for running the equipment and for cooling the facilities to dispel the heat the equipment gives off, with demanding environmental consequences.

Recently, Google and Amazon, two of the biggest, have sought methods for reducing electrical demand and using renewable energy sources. That gives them street cred for being environmentally sound and helps the bottom line by reducing energy costs. At the same time, they are seen as acting ethically.

Do they also have an ethical obligation to push businesses they deal with and have some influence over to act greenly?

E-waste recycling and disposal problems can be reduced by designing products that use fewer toxic components, are easily upgraded so as to delay obsolescence, and are simple to disassemble to separate and retrieve toxic and non-toxic components–that is, value sensitive design. Oversight of producers and recyclers, with appropriate penalties for faulty operations, can alleviate many e-waste problems and keep toxic elements out of the waste stream.

The e-waste problem is a global one, so solutions must be global as well. International pressure on countries that don't cooperate is essential. China, which manufactures so much of the world's electronic equipment and continues to process so much of its e-waste, has a dubious record with regard to green practices. Can international pressure be effective in changing its ways?

Consider also that everything can't be laid at China's doorstep. Indeed, every country that manufactures, consumes, or disposes of electronic devices—that is, every country—must take responsibility for its own actions, must look inward at its practices as well as cooperating in international accords.

And so—reaping what we sow

The vast possibilities of our great future will become realities only if we make ourselves responsible for that future.[23]

Stealing is stealing, whether you use a computer command or a crowbar, and whether you take documents, data or dollars. It is equally harmful to the victim whether you sell what you have stolen or give it away.[24]

Yesterday's science fiction is today's reality and today's science fiction will be tomorrow's reality. But all is not wonderful in this digital dream world, for our marvelous capabilities are as available to malefactors as they are to the ethical and law-abiding. The human psyche is fully capable of leading us to the dark side on occasion, though what that means is not fixed for all time. After all, we are not static creatures. We continually adjust ethical viewpoints as our moral compasses point to changing individual and societal values.

Witness:

- the coarsening of language on cable channel programs and in the movies, where cursing has become so prevalent that it's just a step or two away from being societally normal speech;

- illegal downloading being trivialized as just another harmless aspect of the digital age and thus not deserving of opprobrium;
- the zeal to be the first to post "news" of an event without waiting to check facts, get the full story, or wonder about the impact on someone that erroneous or revealing postings might have;
- the almost addictive need to be in constant communication, even when much of it is trivial and regardless of whether it intrudes on others;
- the rush to buy the latest IT device even though the "old" one is perfectly functional, thus contributing mightily to the e-waste mountain.

Consider these and others as actions, not value judgments, as choices we make, not givens. Perhaps it's all just part of a natural evolution. After all, technology doesn't stand still and neither do we. There is bound to be some swinging back and forth across what are eventually adopted as the lines of societal norms. As technological developments and social values intersect, swings are inevitable, as are shifts in norms.

So, our sense of self, the routines of everyday living, the nature of our communities, the distribution of power in our societies, all are in flux. We've seen dramatic changes in the way individuals and cultures affect and are affected by technologies. Social, political, economic, and business impacts, along with ethical and legal issues all intertwine as parties to those effects. We've witnessed the interplay and seen how different views of ethical behavior judge outcomes positively and negatively.

Who benefits from the digital age, the blossoming of IT, the incredible fingertip ready access to resources, and who doesn't? Is the common good better served if universal access is provided no matter the cost? Is it reasonable to expect the haves to subsidize the have-nots? Can ethical sensibilities and that rare commodity common sense help steer prudent legislation while simultaneously serving as guides to reason that can fill in the gap between slow-moving legislative action and rapidly advancing technologies? Whatever the conclusions, the Internet complicates the picture because, cooperative agreements notwithstanding, nations have no control over what originates in other nations and often not much control over what happens in their own.

What we know for sure is that technological advances will continue at a pace too fast for the time we have to figure out the best course to take with them or to deal with their consequences. At times IT seems to take on a life of its own. Nevertheless, the better we understand its impacts the better we can use IT wisely as a powerful means for serving our more noble instincts.

So the struggle to keep up with IT, use it sensibly and wisely, create meaningful rules and procedures, follow proper use, improve access, and preserve individual rights, goes on. If we keep in mind that acting ethically and taking responsibility for our actions must accompany these quests, there is good reason to be hopeful.

Notes

1 http://www.searchquotes.com/quotes/author/Immanuel_Kant/2/
2 http://thinkexist.com/quotes/alexander_solzhenitsyn/
3 Elena Kagan (1960–) is currently serving as the 112th justice of the US Supreme Court.

4 http://www.marylandindependentparty.org/for_discussion/Kagan_confirmation_hearings_transcript.html

5 http://www.npr.org/blogs/thisisnpr/2011/10/24/141650305/an-update-for-atc-listeners

6 Charles Rangel (1930–), Representative, NY 15th congressional district, and third longest serving member of the House of Representatives.

7 James McGreevey (1957–), 52nd governor of New Jersey, is now studying to become an Episcopal priest.

8 http://www.nj.com/news/ledger/index.ssf?/news/mcgreevey/stories/20040813sl_mcg_quits.html

9 See Chapter 3.

10 http://epic.org/free_speech/cda/

11 http://www.law.cornell.edu/supct/html/02-361.ZS.html

12 Rosalind W. Picard, *Affective Computing*, MIT Press, 2000, p. 135.

13 Isaac Asimov, *Runaround*, Street and Smith, New York, 1942.

14 Isaac Asimov, *Robots and Empire*, Doubleday, 1985. (Reference to Law 0 was added in sequence to the first three laws as exceptions to them.)

15 Lyuben Dilov, *The Trip of Icarus,* 1974.

16 Nikola Kesarovski, *The Fifth Law of Robotics,* in the short story collection *The Fifth Law*, Otechestvo Publishing House, Sofia, 1983.

17 The Association for Computing Machinery (ACM) is "the world's largest educational and scientific computing society, delivers resources that advance computing as a science and a profession" (http://www.acm.org/). Their ethics code is at: http://www.acm.org/about/code-of-ethics

18 The Institute for Electrical and Electronics Engineers (IEEE) "is the world's largest professional association dedicated to advancing technological innovation and excellence for the benefit of humanity" (http://www.ieee.org/about/index.html). The IEEE *de jure* standards organization develops protocol standards for computer hardware and software. Their ethics code is at: http://www.ieee.org/about/corporate/governance/p7-8.html

19 Basel Convention on the Control of Transboundary Movements of Hazardous Wastes and Their Disposal.

20 http://www.ban.org/about/

21 www.greenpeace.org

22 http://www.greenpeace.org/usa/en/campaigns/toxics/hi-tech-highly-toxic/company-report-card/

23 Gifford Pinchot (1865–1946), former governor of Pennsylvania and first Chief of the United States Forest Service, quoted in: http://thinkexist.com/quotations/future/3.html

24 Carmen M. Ortiz, US Attorney General, Massachusetts, describing a federal indictment in *Alleged Hacker Charged With Stealing Over Four Million Documents From MIT Network*, July 19, 2011.

10

The future lies ahead—we're off to see . . .

Human activity is being facilitated, monitored, and analyzed by computer chips in every conceivable device, from automobiles to garbage cans, and by software bots in every conceivable virtual environment, from web surfing to online shopping.[1]

One of the definitions of sanity is the ability to tell real from unreal. Soon we'll need a new definition.[2]

Claims of prognosticators notwithstanding, the future remains shrouded in mystery. Most predictions about noteworthy matters turn out to be wide of the mark; those that have come reasonably close can be attributed as much to good fortune as anything else. Lucky guesses aside, the problem is that predictions depend on some sort of stable pattern, so the ones most likely to be borne out are those for which patterns persist. Quite often, what's most important to predict is change when a pattern no longer holds. That's where, except for inspired speculation, predictions often fail.

Some semi-persistent patterns

When it comes to predicting the future by accurately forecasting magnitudes, time frames, and impacts on us and societies, clarity escapes us. General trends are on somewhat safer ground. Here are a few.

Communication capabilities will grow in scope and become even more vital as the world becomes progressively more connected. That means IT will increase in prominence as a supporting player while at the same time, it blends into the background. Platform convergence will move ahead full speed, but there will always be a place for specialized devices. Clever dedicated apparatuses will find their niches. The use and capabilities of processors ranging from small directly focused application chips embedded in a great variety of items to full-blown computers will continue to expand as their sizes and costs shrink.

Some degree of intelligence, seen and not, controlled by us and not, will be built into almost everything, from clothing to appliances, packaging to products, food to medical treatments, vehicles to houses. IT and computing power, already heading towards utility status, will fade into the background of our consciousness as we take their presence, availability, and performance as given. The cloud utility model of IT service provision will hasten that development. We will be able to utilize great computing power with simple mobile and stationary devices serving as portals to the cloud.

The popularity of social networking will grow, though which sites will falter, which will prosper, and which newcomers will take center stage remains to be seen. Current success, no matter how spectacular, is no guarantee of future viability or influence. That is apparent even in the short lifetime of social networking.

Friendster, launched in 2002, garnered over 3 million users in just three months. MySpace, debuting about a year later, had similar success. Yet both were eventually overwhelmed by Facebook, which arrived on the scene the following year. In another two years, Twitter updated the instant messaging paradigm to huge success. The latter two are only a few years old, but with rare exception, even a few years can be a long time in the digital-techno world, where dominance is often a passing thing.

That can be a paradigm for all technologies, including IT. History is replete with examples of promising, highly rated, and successful technologies and companies that fell by the wayside, were relegated to minor roles or failed outright as others took advantage of business miscues and management shortcomings.

What do you think of this? *Who is right?*

The nature of prediction is such that divergent views are common. Michio Kaku, physicist and professor at City University of New York, claims that Moore's Law, which has held for over 50 years, will "collapse" by about 2020, when we will have reached the limits of silicon technology. Computing power will no longer grow exponentially (http://www.salon.com/2011/03/19/moores_law_ends_excerpt/).

Ray Kurzweil, inventor and scientist, believes that computer power will continue to grow at rates at least as great as Moore's Law. By about 2040 machine intelligence will exceed human's and both will blend in a cyborgian world he calls the singularity. (Ray Kurzweil, *The Singularity Is Near: When Humans Transcend Biology*, Viking, 2005.)

Platform convergence

Platforms are the devices we use to compute, communicate, and connect electronically. Platform convergence refers to any of a variety of devices being able to support all or most of those activities. Television manufacturers are equipping their products with wired and wireless Internet connections, streaming capabilities, and Web browsing. We can use a desktop or laptop computer for productivity software, email, browsing, music and video streaming, watching TV shows, making Internet phone calls, video conferencing, and more. We can use the heading-to-ubiquity smartphones and tablets for many of the same things.

What do you think of this? *Reaching capacity?*

Mobile IT makes use of particular frequencies of the wireless spectrum, a finite resource. Improved compression algorithms and multiplexing techniques can make better use of spectrum capacity, but the spectrum itself cannot be expanded.

> Gartner [estimated] that by the end of 2010, 1.2 billion people [would] carry handsets capable of rich, mobile commerce providing an ideal environment for the convergence of mobility and the Web.
>
> (http://www.gartner.com/it/page.jsp?id=1454221)

The International Telecommunications Union estimate for the end of 2010, which included more basic handsets in the count than Gartner, was "5.3 billion mobile subscribers worldwide (that's 77% of the world population)." In noting that, mobiThinking.com added "90 percent of the world now lives in a place with access to a mobile network" (http://mobithinking.com/mobile-marketing-tools/latest-mobile-stats#subscribers).

If these trends continue, we will reach a spectrum saturation point. What then?

Smartphones have the great advantage of easy mobility, convenience, and access to all sorts of useful and fun apps, but they are not suitable for running productivity software. That's not so much a problem of a limitation of phone power as it is of interfacing. For example, sending a short text message or tweet is a snap, but reading a book on the small screen, working with a spreadsheet, or typing a lengthy document using tiny virtual or physical keys is tedious at best. Some current and future improvements could reduce those limitations.

Video goggles can make the screen look PC size and as they become lighter and more capable, will be a natural accessory to smartphones. When voice recognition and text voicing software improve, documents could be dictated to the phone; books, email, and other text could be spoken by the phone. Commands and requests could be automatically handled—Apple's Siri, their so-called "intelligent personal assistant" for the iPhone 4s, is a first step in that direction.

Lightweight, foldable, easily attached physical keyboards can facilitate using word processing, spreadsheet, and other applications as well. Combined with cloud services, discussed in a following section, all of those activities are feasible even with a relatively low-powered smartphone.

Tablets, a rising presence in the marketplace for their simplicity of use, numerous applications, and portability, somewhat bridge the gap between computers and smartphones. Already popular for playing games, reading digital books, blogs, newspapers, and magazines, their larger screens, larger virtual keyboards, and several ways to go online make emailing, texting, and blogging straightforward. They also will be able to take advantage of cloud services to expand their capabilities.

Dedicated devices have a role   Platform convergence is a trend that is quite likely to continue, but being able to use any device for any purpose will not be the ultimate result. There will continue to be a need and market for specialized and niche products. Some things are just easier and better done with dedicated devices even if particular capabilities can be pushed into generalized devices.

For example, digital photos can be taken with tablets and smartphones and sent to anyone with online or cell network access. Yet digital cameras have many features that would be difficult or ungainly to incorporate into those devices—optical zoom lenses, interchangeable lenses, strong flashes, larger light sensors, and so on. Some or all of those features could be added to tablets and phones, but the additional weight and bulk of the hardware isn't compatible with their design goals.

It's possible to add communication capability to digital cameras so photos could be sent directly, but that would be more awkward than the way tablets and phones work. An automatic short range link between a smartphone or tablet and a camera using a technology like Bluetooth can give us the best of both—take a photo with the camera and send it with the tablet or phone. That's another form of platform integration.

Another example comes from the medical arena. Miniaturized chips with little memory or computing power could be effective at collecting or transmitting the status of illnesses or physical ailments from inside the body. Tracking digestive problems, blood sugar counts, tumor development, nerve signal transmissions, and so on, could be done relatively inexpensively in real time with tiny simple devices dedicated to specific chores that work internally.

Media convergence

Media used to be well differentiated. There were print editions of newspapers, magazines, and books, broadcasts of radio and TV programs, movies and newsreels shown in theaters, recordings of music and stories, all analog in nature. These have come to be called old media. With some minor exceptions like call-in radio programs or letters to the editor, they were one-way technologies staffed and run by field specialists, with content created by accomplished writers, journalists, filmmakers, photographers, and the like.

Digital formats and delivery systems changed the landscape by making it possible for anyone with a connection to create and distribute content on multiple platforms to any one person or to a multitude. That has come to be the meta view of media convergence.

The confluence of platform and media convergence will make anytime anywhere access a major factor in choosing or excluding information sources, whether for serious news, gossip, or entertainment. But instant availability can trump information reliability. Unless a backlash erupts, in-depth reporting will continue to take a back seat to speed. Along with it comes a decline in reliability and further waning of print media.

Still, the providers of old media are not out of the picture. Indeed, despite the death knell being sounded for print media, they not only continue to occupy a space but also produce their offerings in digital form as well, both as duplicates of print pieces and as digital-only offerings.

What do you think of this? *Less is more, less is less*

The Federal Communications Commission (FCC) has long placed constraints on single company ownership of newspapers, TV channels, and radio stations in the same local area, although some waivers have been granted. If those rules are relaxed, as the FCC is proposing, media consolidation will increase. As it is, public interest groups claim that consolidation already has gone too far, with the larger media companies buying up more of the smaller ones, reducing competition, constricting choice, and narrowing access to differing viewpoints.

Proponents of loosening constraints say that media presence on the Internet, reachable any time anywhere, makes more unique content available in addition to repeating what is printed and broadcast. That provides so much greater opportunity for the distribution of all sorts of material that diversity of opinions and competitive reach has actually expanded. Except for the smallest, highly local media, that also makes the concept of a local market passé.

FCC constraints or not, the trend toward media consolidation on many fronts is likely to continue. Is that good news or bad?

Public libraries are keeping pace. More than great stores of physical media that we can borrow or use on-site, they make information available online. Using their Web sites, we can search catalogs, reserve items, and borrow digital material. Many libraries participate in cooperatives and consortia, potentially giving us access to a wide range of material from a single browser session without having to travel anywhere. That trend will continue as systems develop and more libraries join forces.

University libraries are moving in the same direction. Subscribing to digital versions of journals and other professional serials will lessen the pressure on physical shelving space and handling costs. For library patrons, that also means there is no need to queue for items being used by others, although that may be impeded by copyright issues—simultaneous use could be considered illegal copying.

In the digital milieu, media influence is shifting towards IT companies that exist only in digital space. Prime examples are Yahoo, Facebook, LinkedIn, Google, and YouTube (now owned by Google). They have huge numbers of members, readers, and contributors. To establish a foot in both worlds, they are beginning to affiliate with old media content providers.

Mixed media distributors are exemplified by Amazon, which sells video DVDs, and Netflix, which rents them. Both also stream videos to several platforms under various fee structures. Then there are recorded music streaming services, also available on a variety of platforms. Pandora.com, Spotify.com, Grooveshark.com, and Magnatune.com are examples. Another illustration of convergence are sites like Radiorage.com, Live365.com, MikesRadioWorld.com, and Shoutcast.com that stream broadcasts over the Internet from on-air and Internet radio stations.

We're also seeing adoption of a different kind of media convergence—movies featuring actors playing print comic book characters, video games turned into movies and vice versa—so the same content is packaged and repackaged for different media. In itself that sort of convergence is not new—print detective stories and westerns were turned into radio broadcasts, radio shows moved onto TV, and movies were converted to novels. But while the digital world didn't invent that kind of convergence, it certainly has made it easier and more widespread.

The distinction between a content provider and a content distributor is blurring. Cable companies are producing shows of their own as well as carrying content from other producers. Mergers and acquisitions have contributed to convergence. Comcast, a cable content distributor, phone service provider, and ISP, recently bought a 51 percent interest in NBC Universal, a media company that owns television and radio stations, cable channels, movie companies, and theme parks. The Sony Corporation, a manufacturer of digital devices, owns movie, television and music production and distribution businesses, and its game machines and televisions can connect to the Internet.

Social media continue to rise in prominence, gradually encompassing other media forms and offering easy access to a broad swath of participants. That has reached the awareness of public relations firms. Formerly producing one-way messages in support of a company, a brand, a product, or an image, they are shifting focus to tap into the dialogues ricocheting around the Internet within and among social networks and other two-way communications sites. More of that is to be expected.

Cultural convergence

Homophily, literally *love of the same* and figuratively *birds of a feather flock together*, leads us to seek people who are similar to ourselves in various ways. The term also can be applied to seeking congruent information, a homogenizing effect.

One way to cut down on the constant information deluge confronting us is by separating out dissonant data. IT can do that for us, passing what is consonant with our likes, moralities, and world views, and filtering out what is not. We are not always in control of that process, however. For example, the finely directed personalized information and targeted ads that increasingly find their way into our consciousness also constrict what we see.

Of course, there's so much information available to us and so much that bombards us that we do need some sorts of screening mechanisms to avoid chaos or a stifling information overload. Yet in sum, the consequence of filtering

> The public's dilemma is to know how to consume the news with an ability to extract opinion from the simple facts and evidence . . . The best solution . . . is to acquire more diverse information across the ideological and geological divide. If you find yourself relying on one source of information for the news, whether right or left, you are likely to be exposed to more opinion that reinforces rather than challenges your own.
>
> (Nancy Snow, *Information War: American Propaganda, Free Speech and Opinion Control since 9/11*, Seven Stories Press, 2003, p. 31)

is that what we see becomes more homogeneous, narrowing our world view and contracting our awareness.

Ironically, that convergence can lead to divergence as many different subcultures arise, each with its own emphases, agendas, and constituencies. Because of their narrowed viewpoints, they tend not to integrate well with one another. That makes it difficult to achieve consensus on all sorts of controversial issues, and tolerance of other views.

The cloud

In the 1950s and for a couple decades following, computing was the province of businesses, universities, and research institutes running mainframe computers for their primary processing capabilities. Those mysterious machines worked in isolation, located in specially constructed facilities called machine rooms, touched and operated only by trained technicians.

Users submitted their jobs via terminals or by handing in decks of punched cards, and collected their printed or punched output some time later. What happened between those two actions was invisible to the user. Yes, we wrote the programs that ran on those machines and used resident applications as well, but we didn't run them—we just submitted the jobs and retrieved the results. For all practical purposes, mainframes were hidden to us, as if in a cloud.

Fast forward to today's clouds. Though cloud systems are more sophisticated than the relatively straightforward mainframe paradigm of yore, there is a strong resemblance. We connect through the Internet to have mysterious machines do our bidding and the results are returned to us. Now the machines hide in the cloud instead of in the machine room.

Within that extrapolated paradigm, there are many differences. The so-called *public cloud* depends on complex communication networks, most often the Internet, to link users with cloud facilities. That's how we make requests and receive services. Inside the cloud are powerful servers and high capacity storage farms that we never touch or see.

We don't have to write programs, though we can. For the most part, we use prewritten software—word processors, spreadsheets, and database applications, browsers for communication, and special tools for accessing stored content. That content may be what we created, purchased, or whatever the cloud provider makes available. We can own software applications, but we don't have to; the cloud can provide them by what's called *software as a service* (SaaS). In broad strokes, the

Fully configured computers are those with fast processors, a large amount of memory, big hard drives, and several ports for connecting peripherals, including external storage devices. Their opposite, called *thin clients*, are throwbacks to the terminals used in the mainframe scenario. The term typically referred to desktop computers with modest processors and memory, no hard drive, and ports only for a keyboard, monitor, and mouse. From a functional viewpoint the thin client paradigm readily applies to smartphones, tablets, and other online access devices.

public cloud can be thought of as a hosting service that provides software, access to data, and storage, backup and recovery of our own data, and the machinery to run applications and deliver results.

That gives us a great deal of freedom, because we're not restricted to a particular device, say a desktop computer, to utilize the power of the cloud. Furthermore, none of our access devices needs to be fully configured computers. So we can take advantage of platform convergence, or more specifically, platform independence, and location independence—any device capable of connecting to the Internet from wherever we happen to be will open the cloud to us.

Other advantages to public cloud computing are a function of the fact that almost all the support work is done by the provider. Setup is internal to the provider, which also manages the services and associated equipment, including hardware and software upgrades. We only need access devices and appropriate broadband connections. The service scales up and down easily as needs change. Charges are based on what is required. For all those reasons and because businesses using the cloud can eliminate some in-house technology and support staff, cloud computing is gaining a significant foothold in the corporate world.

Alongside the many advantages are inevitable downsides. First, there are several public cloud service providers[3] and they are not interchangeable, so the provider contracted with is the one that has to be used. If the cost structure changes, you either must accept it or change providers. If you do change, you may find it difficult to transfer all your data from one to the other or find all the software you have been using.

Security is a critical issue. Although cloud providers pay close attention to security, we know that incursions can happen. The rich trove of corporate and personally identifiable information held in the cloud can be a tempting target for hackers. Exposure, whether from inside the cloud or by break-ins, could have disastrous consequences. Reluctance to store sensitive information with a cloud provider is understandable.

Service interruptions can loom large. Reliance on the cloud means a forced wait in the face of outages. Then there's the issue of stability. If providers merge, you might be switched to a contract with terms you don't like. Worse, if your provider goes out of business, your data might go too, and the loss of services could cause crises among its many users, especially the corporate ones.

Less dramatic but noteworthy nonetheless, the advantage of having system management and maintenance done by cloud providers can be offset by the fact that you or businesses have no control over what the providers do. The quality and consistency of maintenance and upgrade schedules, how quickly they respond to problems, how readily they match fluctuating needs, how well they meet contract terms, are in their hands. Companies in particular are well advised to monitor cloud services, paying close attention to what's being delivered in comparison to their *service level agreements* (SLAs).[4]

Taking the pros and cons into account, businesses are likely to be reluctant to sign on to a future completely in the public cloud. Even if there are no provider issues, since everything goes through the Internet, connection slowdowns or failure means the service does likewise, causing disruptions that could negatively affect business operations. The most likely scenario is a dual solution—keep mission critical operations and sensitive information in-house and move the rest to the public cloud.

More clouds in the forecast   All things considered, public clouds are becoming more popular. As they stabilize their operations and improve their services, that trend is likely to continue. At the same time the cloud paradigm is so intriguing that businesses are following the model internally, constructing *private clouds*, the local parallel of public clouds. Especially attractive to large corporations for in-house control and handling of critical systems and sensitive data, the idea is leading a shift towards a mix of three technologies: traditional corporate networks, public clouds, and private clouds.

That blended approach, which can take advantage of the best uses of each technology, is likely to grow. As it does, IT departments will be refocused. Keeping traditional and private cloud corporate systems secure and reliable requires a significant support staff that, given the ever changing probes of hackers, must have strong expertise in network protection and system recovery. Routine system support, which can be moved to a public cloud or locally outsourced, will be deemphasized. Providing strong intrusion protection and rapid recovery will become paramount.

What do you think of this? *Cloud internetworks*

Just as businesses outsource some services to public cloud providers, so could the providers outsource to each other. An interconnected network of clouds or a consortium-based cooperative of cloud providers and aggregators, with appropriate redundancies and backups, could improve service reliability and free customers from many of the anxieties that incompatible and standalone providers engender.

Would provider competition forestall those models? Are there regulatory issues?

Revaluing privacy

Before the digital age, privacy was taken as the default condition of an obviously important right. Although we divulged personal information to financial institutions, government agencies, and so on, privacy was assumed to be protected. Opportunities for mass revelations were few and far between.

How to balance privacy, security, anonymity, and the quest for personal information on the Internet is becoming a major problem. It began slowly, almost unnoticed, with the shift from analog to digital systems and from standalone to interconnected networks. The convenience provided by online access started to overtake privacy concerns even as the opportunities for unwanted disclosure increased. The arrival of social networks, with their easy posting and information sharing, led to an explosion of participation before the enormous implications of ready information access were realized. Now we are seeing the beginnings of a pushback.

Identity, hidden or revealed   Facebook, LinkedIn, and newcomer Google+ require participants to use their real names; YouTube, WikiLeaks, and Twitter don't, although Twitter does weed out people trying to use celebrity names as their own. There are good arguments for both sides of the issue.

Anonymous posting opens the door to bad actors. Forcing the use of real names shines a light on posters, making them accountable for what they say. That cuts down on vitriol, false allegations, hate speech, bullying, and many other abusive practices. Yet there are times when being anonymous is an advantage, freeing constrained well-intentioned people to express their views and expose wrongdoings in situations where revealing their true identities could be risky.

Free speech is a long-standing tradition in open societies that neither precludes anonymity nor requires disclosure. Pen names, a well-established means of concealing identity, follows that tradition. Are online pseudonyms any different? Any means of concealing identity can protect people with legitimate concerns or disputes who might otherwise be reluctant or unable to express themselves. A major issue in closed societies, it comes up in open societies as well.

> Protections for anonymous speech are vital to democratic discourse. Allowing dissenters to shield their identities frees them to express critical minority views . . . It thus exemplifies the purpose behind the Bill of Rights and of the First Amendment in particular: to protect unpopular individuals from retaliation . . . at the hand of an intolerant society.
>
> (From a 1995 Supreme Court ruling in *McIntyer v. Ohio Elections Commission*, quoted in: https://www.eff.org/issues/anonymity)

Potential whistleblowers might not hesitate to speak out if they had no fear of being identified until they chose to reveal themselves. Employees who want to articulate controversial opinions without being found out by their employers benefit from selective anonymity. People wishing to join online support groups would be free to discuss matters that they feel are nobody else's business, or could prove embarrassing if connected to actual identities. Individuals with important but unpopular ideas might be averse to expressing them if forced to use their real names. Then there is the ability that hiding under a cloaked identity gives to people in repressive societies to expose abuses and express thoughts that otherwise would result in incarceration or worse.

The choice doesn't have to be between full identity or total anonymity. Not everything we want to post requires cloaking. A balanced position would be that real names are required to register on a site but members could choose to post particular items anonymously. The site could track abusive anonymous posters back to their true identity and block them from future participation. That would preserve the deterrence of requiring real names while allowing appropriate anonymity.

That's a sensible option in open societies, providing a court order based on appropriate criteria would be required to have a site reveal someone's identity.

In closed societies, however, the whim of government officials is sufficient to force uncovering the identity of an anonymous user. Allowing the use of a pseudonym coupled with a real-name registration requirement would be no more than a pretense at providing anonymity protection.

In open societies, workable procedures would have to be developed and made obvious to anyone joining a site. Questions to be answered include: Who would define what constitutes an improper posting or transmission? What constitutes abusive posting? Besides a personal affront, does it include a threat to national security or a potentially criminal activity? How would those terms be defined? Importantly, who would be responsible for detecting such material? Would all online activity be screened?

General monitoring by a site or an agency would fly in the face of privacy and overstep the bounds of freedom of expression. Following the *Digital Millennium Copyright Act* (DMCA) model, individual users could notify a site of possible abuse, after which the site would ask the appropriate authorities to look into the situation. That would include their evaluating the material in reference to defined guidelines before taking any action or seeking a court order to uncover identity. That way, site bias would be taken into account. Since the guidelines would be published along with site rules, posters with controversial or unpopular opinions would be aware of them.

Guidelines of those sorts would have to be federally legislated, but that is not to say that any site has to maintain a neutral position. After all, media have their points of view, any of which could be considered as undue bias by people with differing visions. So sites would be free to add their own posting rules as long as they were made clear at the outset and did not contravene federal law.

Sites would not be required to post information inimical to their stated missions. By making their positions on content and anonymity transparent, potential members could avoid sites that they feel are unduly partial or restrictive and current members could leave sites whose policies they find to be unsatisfactory. People would be able to choose their venues circumspectly.

On the other hand, many feel that any legislation that gives government agencies the ability to exercise control over what sites post and what users look for or force sites to follow procedures that require them to reveal identities or proscribe behaviors is antithetical to the principles of free expression and contrary to the functioning of the Internet. Feelings run high in that regard, which in part explains the uproar over the proposed SOPA and PIPA bills that were being drawn up in Congress.[5]

What do you think of this? *Anonymous sources*

News reporters often quote anonymous informants as sources of information for their stories. That preserves the identity of the sources, provides important facts for the reports, and keeps information flowing to the reporters.

WikiLeaks is based on anonymous sources providing the information it posts. Preserving anonymity protects the identity of the leakers, provides important facts, and keeps the information flowing to the site.

> When sources are anonymous, how do we know they are reliable or if they exist at all? How do we know if the information is factual or made up? We have to rely on the ethics of the reporters in the case of news stories, and the ethics of the site operators with regard to WikiLeaks and other sites.
>
> Are these two situations examples of the same thing, or are there fundamental differences? Does the kind of information provided or where it comes from matter?

Tracking and more   Tracking via cookies, key loggers, GPS-enabled smartphones, and the like have been with us for some time. As discussed in Chapter 4, we are just beginning to get a grip on their true privacy implications. We've seen how reactive legislative initiatives have been largely ill-formed, dealing with the issues only marginally and, at times, with too heavy a hand.

Now, a new frontier is being approached. Information-gathering mechanisms are shrinking to near invisible dimensions. As they do, tracking and other forms of data collection will trend towards greater emphasis and reliance on specialized devices. One example is self-acting, self-powered swarms of extremely small sensors called smart dust. These tiny intelligent motes will be able to act on levels from individual to global. Their capabilities will advance from simple recording of such physical data as temperature, movement, light, sound, and vibration, to understanding language and interpreting events.

A convergence is coming—the combination of extremely miniaturized computing circuitry and power sources harnessed to wireless transmission. Realization of enhanced smart dust, incredibly miniscule intelligent, and swarm based, is on the horizon.

Now about the size of rice grains, mote swarms can collect and transmit volumes of information to each other over distances approaching 1,000 feet, relaying it to central locations or to other swarms. At this stage of development, motes cost about $5 each. Inevitably they will shrink in size—down to that of grains of sand and much smaller—and in cost—to fractions of a penny.

Imagine the possibilities for such beneficial uses as health monitoring, air quality alerts, traffic control, security surveillance, temperature and humidity alarms for cargo shipments, and military intelligence gathering. Imagine the possibilities for warrantless spying, tracking our movements, listening to what we say and what is said to us online and offline, matching where we go with whom we meet, and more, just from undetectable nano-size dust on your clothing, whether we have agreed to it or not.

Though it sounds like science fiction, we and other countries have already started down this road and there is likely to be no turning back. If nothing else, we need to be able to defend against smart dust used for externally mounted incursions.

Legislative lag could burden us with serious practical and ethical issues. Support for development is important, not just financially but also to give researchers free rein to pursue many avenues. Balancing that against responsible manufacturing, distribution, use, and access, will not be easy, but delay will only make it harder. Once the dust is out, sweeping it up will be a major chore.

Regulation alone is not enough. Also needed is a solid, unburdened agency empowered with oversight and control mechanisms over the dust itself to ensure that proper usage and privacy protection are being followed. Individual privacy options must include the ability to decline participation, with appropriate exceptions for criminal activity. This won't be easy to get right, but that shouldn't be an excuse. Poorly formed legislation or none at all will shrink the meaning of privacy to the size of the motes.

Facial recognition software   Work on facial recognition is proceeding along two avenues—software that identifies individuals and software that can distinguish facial characteristics but not identify a person. Both have good applications and both carry ominous overtones, not the least of which are privacy concerns.

Good facial recognition identity software can take the place of passwords. Facebook uses the technology to identify friends in posted photos and add tags to them, easing the task of updating photo albums. Police can use images taken by security cameras to identify criminals.

On the other hand, your images, whether in your postings, in someone else's, or just taken surreptitiously as you go about your day, can be used in ways innocuous, annoying, or invasive—for targeted advertising, for movement tracking, and much more—removing privacy from your control. Anonymous excursions, indeed anonymity itself, can be dropped from the vocabulary. Getting lost in a crowd could become a thing of the past.

Software that attempts to determine your gender and approximate age, though not your identity, is being put to use in active-display billboards and signs that present ads for presumably relevant products as you pass by. That gives crowd watchers a new source of entertainment but may be less than amusing to you. More importantly, as the technology develops, unforeseen uses could prove to be overly intrusive, with no opportunity for you to decline being scanned. Will face masks become de rigueur attire?

Legal safeguards fall short   The *Privacy Act of 1974* places restrictions on the use of federally collected personally identifiable information (PII) without written consent of the person involved. Several exceptions are carved out, mostly related to investigations, law enforcement, and computation of census and labor statistics.

The Act was amended in 1988 by the *Computer Matching and Privacy Act* to extend due process before computerized matching programs using PII could be executed. The amended Act applies only to federal agencies, so non-agency bodies like executive branch departments and the court system are not subject to its provisions. That, plus the carveouts, leave a lot of room for skirting privacy protections.[6]

Several other acts dealing with particular areas of privacy have since been passed, as discussed in Chapter 4. All of them leave significant gaps, especially when dealing with potential exposure in digitally based systems.

We've seen many instances of callous disregard for personal privacy among social network, messaging, and sales sites as well as by mobile apps providers and communications carriers. Most European nations have been far more proactive in protecting privacy than the US. Is it time to follow their lead? Must progress in developing new popular and convenient uses of IT come at the expense of further erosion of privacy?

What do you think of this? *Privacy's primacy*

Most European countries have a history of holding individual privacy in high regard, often overriding other considerations. Following that tradition, the European Commission, the executive body of the European Union, is considering legislation that would throw considerable light on Web site data collection activities and give users jurisdiction over what is collected and how it is handled. In essence, what is being addressed is the balance between an individual's control over personal data and site use of the data for commercial or competitive purposes. Indications are that the Commission will give consumers strong protections and require sites to be transparent about their activities and procedures.

Companies would have to clearly explain what they collect and obtain explicit permission from users (opt-in) before acting on their data. They also would have to permanently delete collected information if requested to do so. Data would be broadly defined to include not just enumeration statistics but also pictures, searches, purchases, and the like. Site members would be free to move their information and postings from one site to another. When they ask a site to deactivate their account, collected information would have to be deleted as well.

Would this kind of legislation hobble Web sites? Would it make them less viable by cutting off potential revenue streams? US-based Web sites have traditionally invoked opt-out as the default condition. Users have to explicitly indicate which activities to disallow. That is much weaker protection than opt-in, which prohibits any activity that a user does not directly agree to.

If the Commission's legislation is successful, would it be a good model for the US to adopt? Would pro-business forces combine to thwart such a move? Is individual privacy regarded more highly in Europe than in the US? Should it be?

Network neutrality, up or down?

There are three major types of players in the Internet milieu: content wells (sites that provide content supplied by whomever—YouTube, Hulu, Netflix, Facebook), content distributors (companies that own, manage, and provide access to the networks that transmit content—Verizon, Comcast, AT&T, Optimum Online), and consumers (who download and otherwise utilize content). Network neutrality concerns the first two directly and the third tangentially.

Content wells want no restrictions placed on Internet distribution channels, whether in terms of capacity, speed, or cost structures for differentiated services. They support their position by asserting that without neutrality, smaller companies and startups will be relegated to the slow lanes, limiting their ability to compete effectively. Further, neutrality prevents discrimination against traffic that a distributor may not like for one reason or another, which could threaten free speech. Besides, the Internet has thrived under its neutrality paradigm—changing that will dampen the economic growth and innovation that has long characterized its development.

Content distributors want to be able to control the use of their networks, limiting or throttling back persistent heavy providers and establishing differential fee structures based on network loads and required performance. They justify that stance by claiming that it is entirely reasonable to charge according to usage requirements. Then too, since networks have finite capacities, slowing down high volume traffic can protect against overloads and prevent low demand transmissions from being pushed out of the way. Furthermore, not being allowed to manage their networks appropriately and charge for enhanced service levels will jeopardize their ability to accomplish infrastructure maintenance and build-outs, capacity growth, access enhancements, and speed increases.

Complicating the picture is the fact that many distributors have become content providers as well, either producing their own or by affiliating with or acquiring content businesses. Neutrality prevents content producing distributors from preferential treatment of their own or their affiliates' transmissions, which would place other providers at a disadvantage. The opposing view is that biased service could be prevented with appropriate legislation apart from mandated neutrality, which unnecessarily throttles the ability of distributors to manage their networks.

Consumers are not directly part of the neutrality debate because, neutrality or not, they are subject to the limits of their connections. Consumers already face a hierarchy of speeds and prices. Dialup, the slowest and cheapest Internet connection, uses regular telephone landlines, as does the slightly more expensive and a little faster DSL. Neither is considered broadband, however. That's the province of cable systems and in some instances satellites, but even within those systems there are slower and faster options with different costs.

On the other hand, when a site's pages are slow to load or a download takes longer than expected, the connection is only one factor and not necessarily the limiting one. The server at the referenced site could be undersized, underpowered, or very busy, thus slowing response. The consumer's computer may be infected with a key logger or other spyware. Hyperactive antivirus software may also be the culprit. Still, should neutrality be abandoned, then connection management may indeed be the proximate cause of slow responses.

The FCC has attempted to step into the controversy by declaring network neutrality to be the required mode of operation. Their latest effort took effect in November 2011, after about a year of congressional oversight clashes. However, the FCC's mandate may be short-lived, since many believe the Commission exceeded its authority. That issue should reach a federal appellate court some time in 2012. The losing party will almost certainly try to move the question to the Supreme Court.

There are two fundamental issues at the heart of the FCC's governance purview. One is whether broadband providers run an information service or a communications system that provides transport services. The other is whether wired[7] and wireless providers and systems should be treated equally.

The fact is, while the Web is a conglomeration of information sources that runs on the Internet, the Internet itself is a digital transport system, not an information system, even though information is what it transports. From that perspective, it operates as a common carrier. That's completely parallel to landline telephone networks, which are classified as communications systems even though they carry information. The same can be said of wireless networks. That leads to the conclusion that broadband providers are running a communications service, whether wired or wireless, and therefore subject to FCC rulings.

Should broadband operators running the Internet's networks be permitted to base transport service on content volume or nature, managing their systems as they deem proper? Should they be classified as common carriers, over which the FCC has regulatory authority rather than as information systems? That's what the courts will be addressing.

If it is determined that the FCC lacks the authority to mandate network neutrality and if it is concluded that neutrality is the preferred mode of operation, then there are three options. Congress could expand the authority of the FCC, or it can create a new agency to regulate broadband providers, or it can pass neutrality legislation itself. The debates are likely to continue for some time, but whatever the result, it will apply only to the US.

The shifting digital divide

Not too long ago, providing a computer and a connection to the Internet was the key to shrinking the digital divide. Still an issue in almost every country, including those that are highly developed, now a simple connection isn't enough. While the divide between access and no access is decreasing, the divide between slow connections and high speed broadband is growing.

It's not just a question of broadband availability—it's also a question of affordability. In the US, champion of free market enterprise, where competition is supposed to rule the day, competition is woefully absent in the broadband provider arena. It is rare that more than one cable company offers broadband service in a given area. The cable operators like to say that they have competition from satellite and landline services, but these are not comparable.

At present, satellite companies can offer connections at the relatively low broadband speeds available in the US and cellular networks are gearing up to nationwide 4G (fourth generation) speeds. But for both, the fact is that wireless spectrum is finite and the nature of wireless itself limits the capacity that can be made available. The combination of increasing demand and faster data rates will bring us to a choice point—capacity or speed? That's why carriers are dropping unlimited data plans and changing their rate structures.

A comprehensive fiber optic national network that reaches into home and office connections combined with wireless is the way to the future of nationwide high speed high capacity broadband. Currently the US Internet backbone and second level networks are fiber optic, but at the ISP level and lines to the end users, there is still a lot of copper cable. There also are many areas that are not cabled at all. High speed wireless coverage is spotty as well.

Cost is a considerable problem. Satellites are expensive to build, launch, and maintain and orbit capacity is not unlimited. Constructing and expanding cable systems, whether copper or fiber optic, is also expensive, labor intensive, and time consuming. Building wireless cellular systems is less costly, but tower locations and landline connection availability are challenging. So companies zealously and jealously guard their infrastructure monopolies and spectrum licenses, thereby freezing out true competition.

The oft-maligned *1996 Telecommunications Act* offers a short-run solution: infrastructure owners would have to lease capacity to other companies at below market rates to leave room for competitive services. However, that doesn't address expanding overall capacity.

Here's another possibility: The same philosophy behind the development of a national high speed highway system beginning in the 1950s applies even more directly to a high speed national broadband system. The funds collected on every telephone and cable service bill now dedicated to building out phone service in sparsely served areas, plus federal funds now used to provide basic Internet access, could be used instead to build out a national broadband infrastructure that would be available to any company providing access and service. The enhanced competition would give us broadband speeds comparable to what's common in countries like South Korea and Japan, and at much lower rates than typify US service. Phone and TV would come along with it and all would be able to reach households and businesses countrywide.

What do you think of this? *Are smartphones the answer?*

Smartphones are growing in popularity and with the advertising hype of 3G and 4G services, are being touted as a great way to offer Internet access to everyone. And it may be the case that building out 3G and 4G service is less expensive than building out cabled systems, especially in sparsely populated areas. But consider this:

Only the phone-to-tower connection of cell phones is wireless—the rest uses cables and the existing wired telephone infrastructure with its limitations. Spectrum constraints and interference severely limit data rates over the wireless segment. Phone calls, emails, and texting are one aspect; Web surfing, downloads, streaming, and business uses are another.

Eventually, as demand for high-capacity-requiring applications grows, bandwidth will become saturated. Then too, wireless service is not as reliable as wired service, nor, because of the nature of wireless, can it be. Currently, customer phone, data, and message plan fees are high and come only in packages that usually don't fit very well with actual usage, though enhanced competition should help with that.

So, are smartphones the answer, just a part of the solution, or for the time being, a fairly narrow compromise?

Another look at AI

The real problem is not whether machines think but whether men do.[8]

Scenarios good, bad, and indifferent have been painted for the future that AI holds for us. Arguments over whether machine intelligence will ever match human intelligence won't be settled until it happens, and even then there will be debates about how to make the comparison.

It's not much of a stretch to envision a time in the foreseeable future when all sorts of activities requiring considerable intelligence will become natural to artificial devices, or when their intelligence equals or exceeds that of humans. For the moment, let's suppose we do reach a point where AI and human intelligence are, for all practical purposes, indistinguishable. Will artificially brained creatures operate just as human brains do? Questions remain.

Are there other dimensions of the mind that still will differentiate us, and if so, will that put us at an advantage or a disadvantage? For example, if intelligent machines are unemotional will they make better decisions than emotional humans? What about self-awareness, esthetics, conscience? Will androids, no matter how sophisticated, understand that they are artificial? Will they appreciate and create art, music, literature? Will they behave with integrity, with honor? Will they come to know love?

We can postulate that anything humans know and feel are functions of brain processes and infer that an artificial brain of sufficient complexity will achieve human-like sentience, with all that implies. We may see a future world with new forms of art, science, music, and literature. Imagine robot companionship and pairings, friendships between humans and androids, and artificial offspring created by machines that design and build others. All in all, an intriguing world with amazing potential.

Alternatively, we can hypothesize that we will be in for more of the ravages that humans have inflicted on each other and the environment. After all, automatons with human-like brains need be no more principled than we are. They could take on the less noble human characteristics of jealousy, suspicion, envy, deviousness, and hate— a dismal prospect. Should that come to pass, we might see even more discrimination, warfare, devastation, and environmental decay than we have now. Learning from humans, robots could build a new criminal underworld.

> Artificial intelligence cannot avoid philosophy. If a computer program is to behave intelligently in the real world, it must be provided with some kind of framework into which to fit particular facts it is told or discovers. This amounts to at least a fragment of philosophy, however naive.
>
> (John McCarthy, "Mathematical Logic in Artificial Intelligence," *Daedalus* 117, no.1 (1988): 297–310)

Perhaps a well-formed robot code of ethics will resolve the downside problems, assuming that smart, possibly free-willed automatons would comply. Human ethical codes may suffice in theory, but practice has shown that convenient disregard is an all too common human trait. The same fate could befall any robot code as well.

And what about relationships between humans and machines? Will we see legislation to set boundaries, just as people have proscribed miscegenation, have created race and gender bus seating rules, caste systems, privileged and underprivileged classes, the idolized and the scorned?

It may turn out that whatever level of intelligence is achieved, androids will never be more than human simulacrums, plodding away unfeelingly and automatically without a care or concern for themselves—a new class of slaves. In fact, we might purposely design them that way.

The cyborg scenario   The human/android dichotomy could be combined in a melding of the two—cyborgs. Part human, part machine, cyborgs could be built to overcome human afflictions and even human limitations. We already replace joints and heart valves. Artificial limbs with the same articulation as human limbs, controlled by the brain of the wearer, will be developed. And what more? Imagine replacing eyes, internal organs, spinal cords, or any other human components with artificial ones.

Let's assume we reach that stage. The question becomes, will we stop at replacement or go on to enhancement? How about telescopic vision, being able to run like a cheetah, able to leap tall buildings in a single bound? Is Robocop[9] only a primitive version of a cyborgian future?

Helping or snooping?   When conversing with someone, we look for clues to meaning beyond actual words. An emergent branch of the field of sentiment analysis aims to computerize that kind of judgment by combining voice recognition and understanding with tonal and speech pattern detection so as to discern intentions. We have in one stroke a new openness, more accurate assessment of testimony, another decline in confidentiality, and a way around the bar against self-incrimination.

Mind reading, once the province of the illusions of magicians, mentalists, and charlatans, has been legitimized by the pursuit to create machines that can read minds. Presently, rudimentary mind reading is possible, but requires the cooperation of the subject and large, expensive equipment. Improvements will likely lead to a point where minds can be easily read without cumbersome equipment. That could benefit those with psychoses and give people accused of misdeeds a way to prove their innocence. But if our minds can be read to gather information we hold to ourselves, without our knowledge or complicity, what then of the expansion of the volumes and kinds of personal data already collected?

Workforce shifts—old trend, new twist   From the days when technology first began to have a significant impact in the workplace, advances typically have been met with dire predictions of machines replacing people accompanied by great job losses. In the last couple of centuries, that has applied almost exclusively to manual laborers. For those who were displaced, the situation was dire.

On a national level, the picture looked much better because each new technology eventually created many more jobs than it eliminated. Of course, eventually wasn't much of a consolation to those without the skills needed for the new jobs. Furthermore, the nature of communities changed, and not always for the better.

The industrial revolution of the eighteenth and nineteenth centuries marked the beginning of a great transition, a period during which machinery increasingly replaced workers in factories and on farms, and electrification, transportation, and communications grew rapidly—profound developments all. Over time, there was a huge growth in productivity in all arenas, leading to a significant increase in population together with a movement from farms to cities, and a pronounced upsurge in mass production.

Of course it did not go so smoothly. While overall per capita income rose, income distribution began to skew more towards the rich. Cities became congested, crime rose, and other social ills confronted the nation. Workers who lost their jobs to machines found it hard to make their way. It's easy for those who came out on the plus side, mostly professionals and business owners, to say that job creation more than offset the losses, but not so much for the laborers who suffered in the short term or for factory workers who toiled long hours in unsafe and inhospitable conditions.

The Luddites

In early nineteenth-century England, skilled textile workers were beginning to be replaced by mechanical looms that could be run by unskilled laborers. By 1811, rage against the machines reached a point where a protest movement sprang up. Calling themselves Luddites, the workers marched, petitioned, destroyed machinery, and burned factories. Though the mechanization of textile production was delayed somewhat, it wasn't halted.

The term Luddite became associated with anyone opposed to technology and is sometimes used as a disparaging term.

A very early backlash was the Luddite movement, but onrushing technology was not to be stopped. That phenomenon has continued into the digital age. Increasingly, computer-driven machinery has taken the place of many skilled laborers. For example, we build tireless industrial robots to spot weld automobile components and spray paint vehicles on assembly lines, replacing many blue collar workers who had manually operated the welding and painting equipment.

As computer intelligence improved, job displacement began to creep up the workforce skill level. Computers running specialized software were employed to perform tasks that were well defined by rules and procedures. That was the impetus behind expert systems, whose use allows people to act in place of more highly trained personnel.

AI has pushed that movement farther up the skill scale. Computer software designed to search stores of information to find specifics and identify patterns is increasingly being used to take the place of well-paid professionals, large numbers of whom would otherwise be employed for those tasks and who could not perform them nearly as well. A case in point is the discovery process, a tedious time-consuming labor-intensive part of lawsuits and trials.

Armies of skilled paralegals and lawyers wade through volumes of disclosure material and pour over court records to find precedents and other supports for plaintiffs or defendants in cases to be adjudicated. Those armies are being replaced by computers that sift huge amounts of information, increasingly available in digital form, much more thoroughly, quickly, and cheaply than can people. Large numbers of well-paid professionals will find themselves displaced by the growing use of indefatigable, inexpensive, progressively intelligent machines.

In another highly skilled field, computers running specialized software already are starting to replace computer chip design engineers. As the software improves and design output is linked to production machinery, much of the process of creating and producing new computer chips could be done with little human participation. That could lead to a future where robots with highly developed AI could design their own improvements and build their own offspring. Once again we could say that in the long run, society as a whole will benefit and once again that's easy to say for those who aren't displaced.

Is the tide turning? Now that a different class of labor—skilled professionals, the so-called white collar workers—is beginning to feel the brunt, will that change legislator and public attitudes? Will those replaced by computers in a growing number of fields find they are morphing into Luddites? That would be quite a vindication for the two centuries old movement.

And so—the global imperative, keeping up with the chip

> In this heady age of rapid technological change, we all struggle to maintain our bearings. The developments that unfold each day in communications and computing can be thrilling and disorienting. One understandable reaction is to wonder: Are these changes good or bad? Should we welcome or fear them? The answer is both.[10]

The individual and global impacts of information technologies cannot be overlooked and must not be underestimated. Most of us in free societies would like openness to be the global imperative. We look to ever-capable IT to help it come about, just as closed societies look to IT to maintain control. Locally, we expect our devices to make life more convenient—whether we're looking for information, communicating with friends, colleagues, and associates, arranging for services, traveling, shopping, banking, and whatever else we do throughout the day.

We've seen how each added convenience has come at a price—giving up a little more privacy and a little more control over what's done with the information that's collected about us. Let's look at some reasonable developments and see where extending them might take us.

Envision having a world of conveniences at your disposal, made possible by automatically synchronized cloud connected devices that fully interact with you. Microsoft's Kinect is a primitive precursor of a future when all sorts of platforms will have sensory interfaces that will understand gestures, facial expressions, eye movements, and spoken natural language. Excellent language capabilities will remove the interface limitations of handheld devices.

Have a question? Just ask. Have a request? Just say so. No need to recall a set of allowable words or to structure your query in any prescribed form. Your requests will be carried out. Your answers will be spoken to you or displayed on a screen. Your expressions, intonations, and physical reactions will indicate whether or not you approve, trust the results, or want confirmation.

No need to write anything down—the device will remember for you. All your other digital devices will automatically sync as soon as they are in range, with no action needed by you. Even more convenient, wherever you are everything will be available to any device through cloud service aggregators. And computing power in the cloud will mean that powerful communications capability will be built into the devices, but powerful computing capability will not need to be.

What do you think of this? *Listen up and speak to me*

Siri, a natural language processor that runs on Apple's iPhone 4S, evinces the beginning of the trend for digital devices to respond to spoken queries and commands. Originally sold in the App Store, Siri was destined to be made available for many different mobile platforms. However, since the developers and their patents were bought out by Apple, it will not be available for any non-Apple platforms.

Is that smart competitive action on Apple's part, or simply a way for it to use its purchasing power to stifle development and broader dispersion of a convenience to a larger audience? Will its unavailability to other platform producers stir them to greater innovation or will they take the simpler route of licensing the software should Apple choose to offer that option? Will natural language software development spread or will it become one more victim of the patent wars?

At work, sensors will recognize you, admit you to your workplace and appropriate offices and areas. At home your in-house digital equipment will rouse with you, display the day's events and call up any information you need for the day. Wherever you are working from—at home, in your workplace, or elsewhere, corporate data will be as smoothly available to you as if you were on the premises. You will be able to interact with your colleagues wherever they are, by voice, by video, and by data exchange. At a meeting, whatever you need to display will be brought up automatically and distributed electronically. You may be the only person in the room, as may the other attendees be in theirs, but everyone will see and hear each other quite plainly in real time.

Your home's entertainment gear, heating and cooling systems, appliances and utilities, all sensor equipped, will be similarly responsive. Then add full-blown automatons and life couldn't be better—or could it? Underlying this technologically supported nirvana are the assumptions that the power grid and communications networks function flawlessly, that all software is bug free, that competition doesn't prevent cross-platform and cross-service compatibility, and that communications capacity and speed can keep pace. Would a systems failure leave us immobile?

While we may never reach the idyllic state, elements of it are sure to arrive. Sensory interactivity, enhanced device capabilities, and automatic information exchange is on the horizon. But as we've seen over and over again, the downside is never out of the picture. Whatever level of convenience we achieve, careful diligence will be paramount. If the past is any guide, the default setting on all our devices will be to automatically exchange data with any device within range. Unless specifically constrained, information will be transmitted and available to every third-party application provider and distribution channel; it will not be confined to the devices. Opportunities for snooping will abound, tempting rogues and official agencies to cut corners. With so much tantalizing data waiting to be seized, hackers will flock to the challenge.

The potential for intrusions will require a substantial escalation of the capabilities of digital safeguards and unrelenting diligence in using them. Security application developers will see their industry grow substantially.

Legislation at the national level aimed at keeping a sensible rein on abuses is essential. But the speed of technological change implies that the historically slow, reactive pattern followed by our legislators won't do. While it's true that the full impact of new IT, or any new technology for that matter, isn't known until it happens, sensible expert-based outcomes can be reasonably expected and legislatively accounted for beforehand.

Sole reliance on proliferating and expanding congressional committees and government agencies is not the best practice. Instead, field experts can be engaged, especially those involved in developing the IT, along with consumers most likely to be affected. Standard issues to be included from the start are the impacts on privacy, confidentiality, disclosure, and control of PII.

The reactive mode can be reserved for adjustments to account for gaps or the unexpected. Of course, legislative lag will always be with us, but we can hope that irrelevant partisan disputes and bickering that take precedence over practical solutions can be dispensed with, so as not to delay creating appropriate laws.

Legislation must be accompanied by provision for careful oversight to assure compliance. Effectiveness dictates that authorities and agencies at all levels should be cooperatively engaged. We've seen how privacy violations crop up with every new IT application and again as a site's activities are changed or expanded. There's no reason to believe that won't continue to happen. We've also seen how zealous enforcement can itself lead to abuses, so the balancing acts will continue.

Internationally we know that one nation's laws do not have automatic standing in any other nation, nor do they necessarily sway their leaders, legislators, and judges, so in addition to national efforts we have to move on international agreements. The worldwide community of Internet users is a global society with all the factions, fractures, special interests, and competitors jousting for influence, if not dominance, of any society. Can we capitalize on that diversity to morph it into an international commons we all participate in, or will it lose its marvelous universality and essential independence?

Now more than ever, globally supported IT makes international cooperation paramount. That will require radical change from historical patterns. Otherwise, international cooperation will remain nothing more than an unrealized vision.

At this point, whether current and developing information technologies will be used to gain the upper edge in battles over uncompromisable goals, or will give us the opportunity to realize a saner more peaceful future, is a toss-up. The optimistic view is that a balance can be found. Are we players in that future or is it out of our hands?

Regardless, we always will have choices to make and we will have to live with the consequences of those choices. So in the final analysis, as we dive into the bitstream it's up to each of us to keep our heads above the flow and swim in the right direction, divisive currents notwithstanding.

Notes

1   Wendell Wallach and Colin Allen, *Moral Machines: Teaching Robots Right from Wrong*, Oxford University Press, 2008.
2   Alvin Toffler: http://www.quotegarden.com/society.html
3   IBM, Amazon, Google, and Microsoft are currently among the larger public cloud service providers.
4   Service Level Agreements are contracts with providers. They specify the services to be delivered and their levels, the metrics used to measure the *Quality of Service* (QoS) provided, and the cost structure.
5   See Chapter 5 for a discussion of the events and their outcomes.
6   The text of the amended Act is at: http://www.justice.gov/opcl/privstat.htm
7   Wired includes both electrical (copper) media and light-based (fiber optic) media.
8   B.F. Skinner, *Contingencies of Reinforcement: A Theoretical Analysis*, Appleton-Century-Crofts, 1969.
9   *Robocop*, dir. Paul Verhoeven, Orion Pictures, 1987. A mortally wounded cop is restored as a powerful cyborg and returns to the police force.
10 http://www.technorealism.org/

Addendum

Website popularity statistics, though only a snapshot, can paint an interesting picture of how we use online sources of news, media, and politics, conduct e-business, and generally surf. Here are the top 10 sites in a few of the categories compiled by *ebizmba* as of December 1, 2011 (http://www.ebizmba.com/articles/news-websites). These are revealing not only for which sites make the lists, but also for the comparative visitor volumes within and among the categories. (Parentheses show the estimated number of unique monthly visitors in millions.)

News Web sites

1. Yahoo! News (110)
2. CNN (74)
3. MSNBC (73)
4. Google News (65)
5. New York Times (59.5)
6. Huffington Post (54)
7. Fox News (32)
8. Digg (25.1)
9. Washington Post (25)
10. LA Times (24.9)

Reference Web sites

1. WikiPedia (350)
2. Yahoo Answers (85)
3. About (78)
4. Answers (57)
5. eHow (55)
6. Reference (40)
7. HubPages (28.5)
8. Squidoo (24.8)
9. Google books (24)
10. wikia (23.1)

Media Web sites

1. EW (9)
2. Hollywood Reporter (4.5)
3. Deadline (1.3)
4. MediaBistro (1.25)
5. MediaITE (.95)
6. TheWrap (.55)
7. Variety (.525)
8. AdAge (.5)
9. MediaPost (.4)
10. PaidContent (.35)

E-business Web sites

1. Mashable (7)
2. TechCrunch (6.5)
3. DigitalPoint (4.45)
4. ZDNet (4.25)
5. WebmasterWorld (4.2)
6. Arstechnica (3)
7. Sitepoint (2.9)
8. ReadWriteWeb (2.7)
9. cnet (2.3)
10. Internet (2.1)

Political Web sites

1. Huffington Post (54)
2. Drudge Report (14)
3. Politico (5)
4. Salon (4.3)
5. Newsmax (4.2)
6. TheBlaze (4.1)
7. CSMonitor (4)
8. Washington Times (2.5)
9. RealClearPolitics (2)
10. TheHill (1.7)

Social media Web sites

1. Facebook (700)
2. Twitter (200)
3. LinkedIn (100)
4. MySpace (80.5)

 5. Ning (60)
 6. Google+ (32)
 7. Tagged (25)
 8. Orkut (15.5)
 9. Hi5 (11.5)
10. myyearbook (7.45)

Additional reading

Hal Abelson, Ken Ledeen, and Harry Lewis, *Blown to Bits: Your Life, Liberty, and Happiness After the Digital Explosion*, Addison-Wesley Professional, 2008 (and creative commons license).

Peter Andreas and Kelly M. Greenhill (eds), *Sex, Drugs and Body Counts: The Politics of Numbers in Global Crime and Conflict*, Cornell University Press, 2010.

Sara Baase, *A Gift of Fire: Social, Legal, and Ethical Issues for Computing and the Internet*, 2nd ed., Prentice-Hall, 2008.

Gordon Bell and Jim Gemmell, *Total Recall: How the E-Memory Revolution Will Change Everything*, Dutton Adult, 2009.

Mira Burri-nenova (author) and Christoph Beat Graber (editor), *Intellectual Property and Traditional Cultural Expressions in a Digital Environment*, Edward Elgar, 2008.

Nicholas Carr, *The Big Switch: Rewiring the World, from Edison to Google*, W.W. Norton, 2009.

Ronald Deibert, John Palfrey, Rafal Rohozinski, and Jonathan Zittrain (eds), *Access Denied: The Practice and Policy of Global Internet Filtering*, MIT Press, 2008.

Cory Doctorow, *Content: Selected Essays on Technology, Creativity, Copyright, and the Future of the Future*, Tachyon Publications, 2008.

Allison H. Fine, *Momentum: Igniting Social Change in the Connected Age*, John Wiley, 2006.

David D. Friedman, *Future Imperfect: Technology and Freedom in an Uncertain World*, Cambridge University Press, 2008.

Bob Garfield, *The Chaos Scenario*, Stielstra Publishing, 2009.

Harry J. Gensler, *Ethics: A Contemporary Introduction*, Routledge, 2011.

Jack Goldsmith and Tim Wu, *Who Controls the Internet: Illusions of a Borderless World*, Oxford University Press, 2006.

David Holtzman, *Privacy Lost: How Technology Is Endangering Your Privacy*, Jossey-Bass, 2006.

Sheena Iyengar, *The Art of Choosing*, Twelve, 2010.

Deborah Johnson and Jameson Wetmore (eds), *Technology and Society: Building Our Sociotechnical Future*, MIT Press, 2008.

Josephine M. Kerr and Rebecca E. Johnstone (eds), *Networking and Information Technology: Designing a Digital Future for the U.S.*, Nova Science Publishers, 2011.

Joseph Migga Kizza, *Computer Network Security and Cyber Ethics*, McFarland, 2011.

Lawrence Lessig, *Free Culture: How Big Media Uses Technology and the Law to Lock Down Culture and Control Creativity*, Penguin, 2004.

Farhad Manjoo, *True Enough: Learning to Live in a Post-Fact Society*, Wiley, 2008.

Miguel Nicolelis, *Beyond Boundaries: The New Neuroscience of Connecting Brains with Machines—and How It Will Change Our Lives*, Times Books, 2011.

President's Council of Advisors on Science and Technology, *Designing a Digital Future: Federally Funded Research and Development in Networking and Information Technology*, Books LLC, Reference Series, 2011.

Michael J. Quinn, *Ethics for the Information Age*, 4th ed., Addison-Wesley, 2010.

Byron Reeves and J. Leighton Read, *Total Engagement: Using Games and Virtual Worlds to Change the Way People Work and Businesses Compete*, Harvard Business Press, 2009.

Emma Rooksby and John Weckert, *Information Technology and Social Justice*, IGI Global, 2006.

Scott Rosenberg, *Say Everything: How Blogging Began, What It's Become, and Why It Matters*, Crown, 2009.

Stanley Schmidt, *The Coming Convergence: Surprising Ways Diverse Technologies Interact to Shape Our World and Change the Future*, Prometheus, 2008.

Clay Shirky, *Here Comes Everybody: The Power of Organizing without Organizing*, Penguin Press, 2008.

Lee Siegel, *Against the Machine: Being Human in the Age of the Electronic Mob*, Spiegel & Grau, 2008.

Edward Tenner, *Why Things Bite Back: Technology and the Revenge of Unintended Consequences*, Alfred A. Knopf, 1996.

Fred Turner, *From Counterculture to Cyberculture: Stewart Brand, the Whole Earth Network, and the Rise of Digital Utopianism*, University of Chicago Press, 2008.

Joseph Turow and Lokman Tsui (eds), *The Hyperlinked Society: Questioning Connections in the Digital Age*, University of Michigan Press, 2008.

Siva Vaidhynathan, *Copyrights and Copywrongs: The Rise of Intellectual Property and How It Threatens Creativity*, New York University Press, 2001.

David Weinberger, *Everything Is Miscellaneous: The Power of the New Digital Disorder*, Times Books, 2007.

Jonathan Zittrain, *The Future of the Internet and How to Stop It*, Yale University Press, 2008.

Index